Lecture Notes on

INTRODUCTORY THEORETICAL ASTROPHYSICS

ASTRONOMY AND ASTROPHYSICS SERIES

Editor: A. G. Pacholczyk

Basic Physics of Stellar Atmospheres
T. L. Swihart

Physics of Stellar Interiors
T. L. Swihart

Lecture Notes on Introductory Theoretical Astrophysics
R. J. Weymann, T. L. Swihart, R. E. Williams, W. J. Cocke,
A. G. Pacholczyk and J. E. Felten

In preparation:

A Handbook of Quasistellar and BL Lacertae Objects
E. R. Craine

Planetary Interiors
V. N. Zharkov, V. P. Trubitsyn and L. V. Samsonenko
Edited by W. B. Hubbard

A Handbook of Radio Sources
A. G. Pacholczyk and M. Tarenghi

The Relativity of Space-Time
D. J. Raine and M. Heller

Lecture Notes on

Introductory

Theoretical Astrophysics

R. J. Weymann, T. L. Swihart
R. E. Williams, W. J. Cocke
A. G. Pacholczyk, J. E. Felten

Steward Observatory, The University of Arizona

Pachart Publishing House

Tucson

Library of Congress Catalog Card Number: 75-19595
International Standard Book Number 0-912918-14-4

Pachart Publishing House
Post Office Box 6721, Tucson, Arizona, 85733

Contents

Preface

Experience at the University of Arizona has indicated that today's graduate students entering an Astronomy Department seeking a Ph.D. in Astrophysics are a fairly diverse group. While many still will have had substantial exposure to basic Astronomical and Astrophysical concepts while undergraduates, an increasing number are undergraduate Physics majors with very little or no contact with Astronomy or Astrophysics. To provide a bare minimum background in Astronomy and Astrophysics, we have found it desirable to introduce four one-semester Introductory courses in Stellar Astronomy, Theoretical Astrophysics, Instrumentation, and Solar System Astronomy which nearly all entering graduate students take. The present Lecture Notes on Introductory Theoretical Astrophysics are an outgrowth of the course in Theoretical Astrophysics.

Obviously, it is ludicrous to attempt to cover, even in a very cursory manner, the whole of what we today could reasonably include in the term Theoretical Astrophysics. We have therefore chosen to select (naturally enough) those topics of most interest to the Arizona Faculty who participated in the teaching of the course. Even so, it is obvious that to digest even introductory material covering Radiative Transfer and Stellar Atmospheres, Stellar Evolution and Nuclear Astrophysics, Gaseous Nebulae and Interstellar Matter, Relativity in Astrophysics, Theoretical Radio Astronomy and Plasma Physics, and High Energy Astrophysics is asking quite a bit in one semester for a first year graduate student. We believe there is ample material for a full-year course.

Nevertheless, our experience has been that even one semester course succeeds in at least exposing the student to some of the current content and basic concepts of Theoretical Astrophysics, and allows him to proceed directly to an advanced course (which should include some directed research) in any of these areas and this has been our goal.

A word on style and uniformity is in order. These notes evolved from an experiment in "team teaching" in which several faculty members participated in the course every semester. This has some obvious disadvantages, but has some advantages too, not the least of which is the fact that entering students quickly become acquainted with many of the faculty and vice versa. As a consequence of this, however, the style and level of presentation of each portion is quite different. We have made no attempt to make it uniform, except to attempt to make the order of presentation and notation consistent.

We hope others outside of the University of Arizona will find these notes useful in teaching and learning the fundamentals of modern Theoretical Astrophysics.

R. J. WEYMANN

The University of Arizona
Tucson, Arizona
September 1975

1.

Radiative Transfer and Stellar Atmospheres

1.1 Specification of the Radiation Field

Practically all the information about extraterrestrial objects comes to us in the form of electromagnetic radiation, whether it be gamma rays, x-rays, visible lights infrared, or radio waves. Thus it is necessary for all astronomers to have some familiarity with the interaction of matter and radiation and of the essentials of the theory of radiative transfer.

A completely correct description of the electromagnetic field and its interaction with matter demands a quantization of the electromagnetic field (see for example Leighton, Chapter 6). Depending upon the circumstances, however, the electromagnetic field can often be considered on the one hand in terms of the completely classical sense of Maxwell's equations--electromagnetic waves whose sources are charges and currents--and on the other hand in terms of a set of particles of definite energy (photons) which are created and destroyed upon interaction with matter. For our purposes, the latter description is adequate and is the simplest, conceptually.

How do we give a mathematical description of the swarm of photons? Let us first recall the distribution function, f, familiar to students of kinetic theory:

$$\Delta N = f(x,y,z,P_x,P_y,P_z,t)(\Delta x\Delta y\Delta z)(\Delta P_x\Delta P_y\Delta P_z) \tag{1.1}$$

is the number of particles located at some instant t in a volume of space $\Delta V = \Delta x\Delta y\Delta z$ centered at x,y,z whose momenta lie inside a volume of momentum space $\Delta V_p = \Delta P_x\Delta P_y\Delta P_z$, centered at P_x,P_y,P_z. Of course, one could use cylindrical or spherical co-ordinates to describe either the space-dependence or momentum dependence of f.

If we use spherical co-ordinates to describe the momentum dependence

$$\Delta P_x\Delta P_y\Delta P_z = P^2\sin\theta_p\Delta\theta_p\Delta\phi_p\Delta p = P^2\Delta\Omega_p\,\Delta p, \tag{1.2}$$

where $\Delta\Omega_p$ is a small solid angle in momentum space. In all the cases we shall consider, there is assumed to be plane symmetry as far as the spatial dependence is concerned (f does not depend upon y,z), axial symmetry as far as the momentum is concerned (f does not depend upon ϕ_p), and a steady state (f does not depend upon t). Thus:

$$\Delta N = f(x,\ \theta_p,P)P^2\ \Delta\Omega_p\ \Delta p\ \Delta V. \tag{1.3}$$

In radiative transfer, it is customary to describe the radiation field, not in terms of the distribution function f, but rather in terms of a function closely connected to f called the specific intensity $I_\nu\ (x,\theta\ )$. I is defined in terms of the energy transported in time Δt by photons at position x across a small surface of cross section ΔA (whose normal lies in some direction θ), the photons having frequencies in the range ν to $\nu + \Delta\nu$ and directions confined to $\Delta\Omega$:

$$\Delta E = I_\nu\ (x,\theta)\Delta A\Delta t\Delta\Omega_p\Delta\nu. \tag{1.4}$$

The connection between f and I can be seen as follows. All the photons destined to cross the surface ΔA in time Δt are to be found immediately behind the surface in a small volume $\Delta V = \Delta Ac\Delta t$ which happen to be headed in directions contained in solid angle $\Delta\Omega$.

The momentum increment Δp corresponding to frequency interval $\Delta\nu$ is $\Delta p = \frac{h}{c}\,\Delta\nu$. Since each photon carries with it an energy $h\nu$, from equation (1.3) we have

$$\Delta E = (h\nu)\Delta N = (h\nu)fp^2[\Delta Ac\Delta t]\Delta\Omega_p\,\frac{h}{c}\,\Delta\nu. \tag{1.5}$$

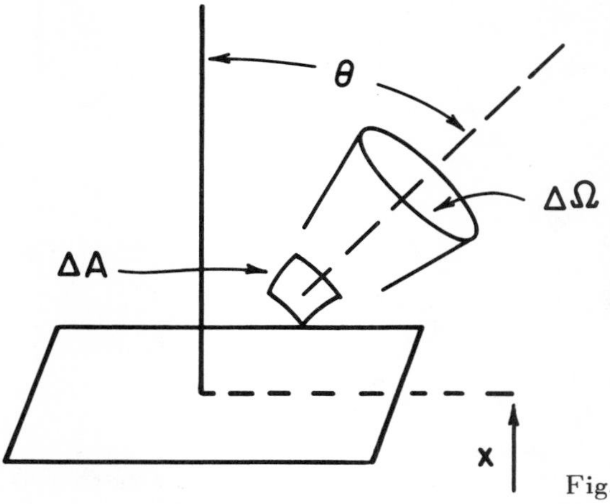

Fig. 1.1 Defining specific intensity

Equating this to the same energy transported in terms of equation (1.4) we see that f and I_ν are simply related by

$$I_\nu = \left(\frac{h^4 \ \nu^3}{c^2} \right) f. \tag{1.6}$$

In quantum statistical mechanics, the occupation index, η is the number of particles in a cell of phase space of volume h^3, $\Delta\tilde{N} = fh^3 = \eta$, hence

$$I_\nu = \frac{h\nu^3}{c^2} \eta . \tag{1.7}$$

(Of course, the precise definition of all quantities like I_ν really involves the limit of $\Delta E \ / \ (\Delta t \ \Delta\nu \ \Delta A \ \Delta\Omega)$ as Δt, $\Delta\nu$, ΔA, $\Delta\Omega$ all approach zero). The specific intensity does not fully describe the radiation field because the field may be partly polarized; in that case four quantities (the four Stokes parameters) are required to completely describe the radiation field, rather than simply I_ν (see Chandresekhar, Chapters 1-15). It is important to realize that in the definition of I_ν in equation (1.4)(cf. Figure 1.1) the area ΔA is normal to the direction of photon flow considered. Alternatively, it is possible to consider a surface of area ΔA with photons flowing across it at an angle θ. Arguments essentially the same as these used to establish the relation between f and I_ν lead us to the conclusion that in this case

$$\Delta E = I_\nu(x,\theta)(\Delta A \cos\theta)\Delta t \Delta\nu \Delta\Omega, \tag{1.8}$$

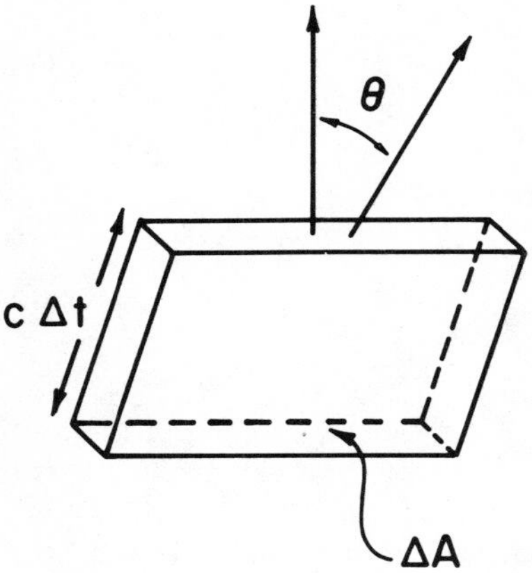

Fig. 1.2 Volume containing photons

since the volume containing the photons is not $\Delta Ac\ \Delta t$ but $\Delta Ac\Delta t \cos\theta$ (cf. Figure 1.2).

The flux, a quantity of considerable interest in astrophysics, is the net rate at which energy is being transported across a fixed surface by all the photons, regardless of their direction. Since the surface is fixed we use equation (1.8), and we must evidently integrate over all directions to find the energy transported across a surface of area ΔA in time Δt by photons of frequency in the range $\nu, \nu + \Delta\nu$. This quantity is called the monochromatic flux $F_\nu(x,\nu)$. If the surface area is taken to be in the plane of symmetry of the radiation field,

$$F_\nu(x) = \iint I_\nu(x,\theta) \cos\theta \, d\Omega = \iint I_\nu(x,\theta) \cos\theta \sin\theta \, d\theta d\phi = 2\pi \int_0^\pi I_\nu \cos\theta \sin\theta \, d\theta = 2\pi \int_{-1}^{+1} I_\nu(x,\mu)\mu d\mu \ , \quad (1.9)$$

where $\mu = \cos\theta$. Obviously the flux associated with all photons, regardless of their frequency, is given by

$$F = \int_0^\infty F_\nu d\nu. \quad (1.10)$$

A second quantity of interest is the energy density, U. This is clearly just the total number of photons per unit volume, N, multiplied by the energy of each photon. The total photon density is obtained by integrating the photon distribution function over all momentum space

$$N = \int f p^2 d\Omega dp = \int \left[\frac{I_\nu d\Omega}{ch\nu} \right] d\nu \; ; \tag{1.11}$$

$$U = \int (h\nu) f p^2 d\Omega dp = \int \frac{I_\nu}{c} d\Omega d\nu \; . \tag{1.12}$$

The <u>monochromatic energy density</u> is then just

$$U_\nu(x) = \frac{1}{c} \int I_\nu d\Omega = \frac{2\pi}{c} \int_{-1}^{1} I_\nu(x,\mu) d\mu \; . \tag{1.13}$$

Pressure is defined as the rate of transport of momentum. The expression for radiation pressure $P(x)$ will thus be similar to that involving the flux with the following two differences: (a) the momentum is $h\nu/c$, so an extra factor of $1/c$ will appear; (b) momentum is a vector quantity, hence we must take its components. If we are interested in the component normal to the surface, this will introduce an additional factor of $\cos\theta$.

Hence the expression for the radiation pressure (more precisely, the rate of transport of the normal component of momentum across a surface) is

$$P(x) = \frac{1}{c} \int_0^\infty I_\nu(x,\theta) \cos^2\theta d\Omega d\nu = \frac{2\pi}{c} \int_0^\infty \int_{-1}^{1} I_\nu(x,\mu) \mu^2 d\mu d\nu . \tag{1.14}$$

Note that the three quantities U, F, P involve integrals over μ of the form

$$M_n(x) = \frac{1}{2} \int_0^\infty \int_{-1}^{1} I_\nu(x,\mu) \mu^n d\mu \; , \tag{1.15}$$

where $n = 0,\ 1,\ 2$ for the three cases considered above. The zero-th moment is simply the average value of I, averaged over all directions (the factor 1/2 is simply for purposes of normalization), and is called the mean intensity. It is customary to use the symbols J, H and K to stand for M_0, M_1 and M_2. Evidently

$$U = \frac{4\pi}{c} J, \quad F = 4\pi H, \quad P = \frac{4\pi}{c} K. \tag{1.16}$$

The higher moments ($n \geq 3$) are occasionally used, but have no especially important physical significance. The equivalent monochromatic quantities subscripted with ν, such as $M_{\nu n}(x)$, can be defined by eliminating the integral involving ν in equations (1.14) and (1.15) in the same manner as in (1.12) compared to (1.13).

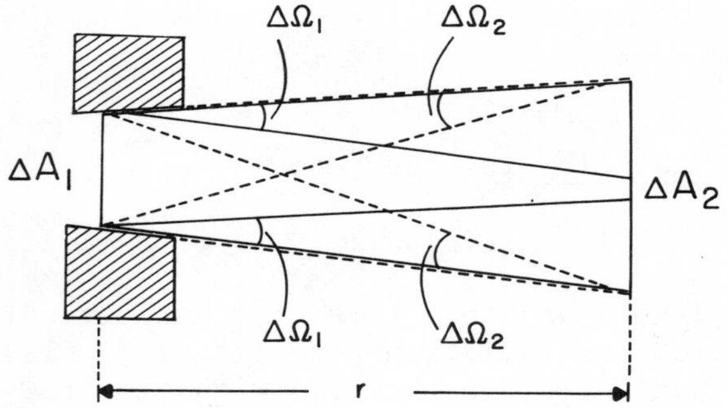

Fig. 1.3 Constancy of specific intensity in empty space

1.2 Interaction of Matter and Radiation

It might at first sight appear that as we followed a particular swarm of photons through empty space, the specific intensity would decrease as the photon swarm spread out and occupied an increasingly larger volume. Evidently, as we follow the swarm the actual number of photons inside the original volume of phase space won't change. If the volume of phase space doesn't change, then evidently f, and hence I_ν must be constant. The fact that the volume of phase space doesn't change is a general result of statistical mechanics known as Liouville's theorem (see ter Haar) although application to a swarm of relativistic particles involves some subtleties. A more direct appreciation of why I_ν is constant as we follow a swarm of photons through empty space can be seen with the aid of Figure 1.3. Figure 1.3 shows a stream of photons passing through surface ΔA_1 and all confined to direction within solid angle $\Delta\Omega_1$. Area ΔA_2 at a distance r from ΔA_1 is constructed so that all those photons (and only those photons) passing through ΔA_1, pass through ΔA_2. Therefore, since $\Delta A_1/\Delta\Omega_2 = \Delta A_2/\Delta\Omega_1 = r$, we have from equation (1.4) that $(I_\nu)_1 = (I_\nu)_2$. This result follows from the reciprocal relation between ΔA_1, ΔA_2, $\Delta\Omega_1$, $\Delta\Omega_2$ as well as the invariance of Δt and $\Delta\nu$. (It is interesting to note that in more general situations, e.g. in the case of the expanding universe, Figure 1.3 can still be used, but the above reciprocal relation must be suitably modified and the result is that I_ν is not constant.)

If we let s be the distance along the direction of travel of the particular photons of interest then the foregoing gives us the most elementary form of the <u>transfer equation</u>, valid in a vacuum:

$$\frac{dI_\nu(s)}{ds} = 0. \tag{1.17}$$

(We will use "s" to denote the distance along any arbitrary ray and will reserve x for distances normal to the plane-parallel atmospheres).

In the presence of matter, however, this equation must be modified as a result of interactions between the photon swarm and the matter. A stream of photons passing through matter will interact with the matter in several possible ways. For example, the photon may excite an atom in the ground state to some excited state, whereupon it will re-emit a new photon of essentially the same frequency, but in a new direction; similarly, a photon may interact with a free electron and be re-directed in a new direction. Such processes are called <u>scattering</u>. On the other hand, a photon may ionize an atom, with the energy of the photon being entirely converted to potential and kinetic energy; similarly the photon may strike a dust grain and its energy entirely converted to internal energy of the grain. These latter processes, in which the "same" photon does not immediately re-appear, are called <u>absorption.</u>

In either case, the original stream of photons is attenuated. (Other, "fresh" photons may join the original stream due to various <u>emission</u> processes, but for the moment we are only interested in the original stream).

In most astrophysical problems we have to deal with a gas in which the interaction takes place independently of the other particles. If we consider a very thin slab of matter of thickness Δs and density n, the probability Δp that a <u>given single photon</u> will interact with the matter in the slab, is simply proportional to the density of particles and the thickness of the slab. The total number of photons interacting with the matter is thus proportional to the number of photons entering the slab, hence

$$\Delta I = -In\,\sigma\Delta s. \tag{1.18}$$

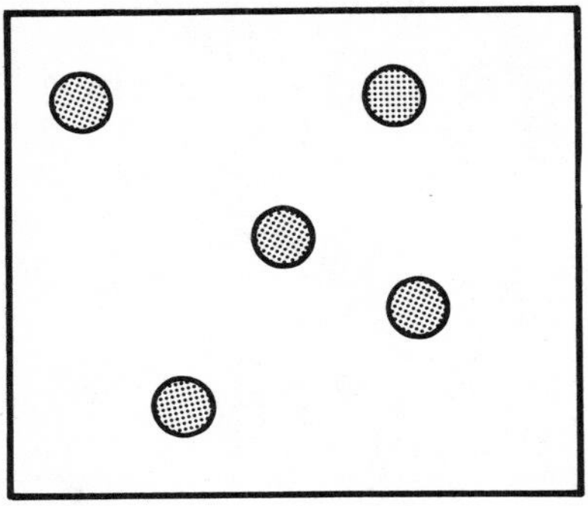

Fig. 1.4 Cross section of ions seen by photons

The minus sign is due to the fact that the beam is decreased in intensity. The constant of proportionality, σ, has the dimensions of area, and is known as the <u>absorption cross section</u>, because of the following simple interpretation of the fraction $|\Delta I/I| = n\sigma\Delta s$ (see Figure 1.4). Imagine the particles to be hard spheres of cross section σ as far as interaction with photons is concerned. In a slab of area A and thickness Δs, there will be a total of $n\Delta sA$ particles and a total area obscured by the particles (assumed not to overlap since Δs is so small) of $An\Delta s\sigma$. The fraction of the area obscured is then just $An\,\Delta s\,\sigma/A = n\Delta s\sigma$. The cross section will in general depend upon the frequency of the photon and will be different for every quantum state of every ion. Its value must be computed quantum mechanically. The density of particles n refers to the number in that particular quantum state and of that particular ion species.

The quantity $n\sigma$ is called the <u>linear</u> absorption and has the dimensions of cm^{-1}. Some photon interactions involve pairs of particles, in which case the concept of cross-section is not especially useful, though one can still speak in terms of a linear absorption coefficient α, with $n_a n_b \alpha$ being the linear absorption. Of course to get the total linear absorption at some given frequency one must sum over the linear absorption associated with each type of interaction.

Consider first the simple case where n is a constant. Equation (1.18) can then be integrated to give

$$I = I_o \exp(-\,n\sigma s), \tag{1.19}$$

or in probabilistic terms, $q(s) = \exp(-n\sigma s)$ is the probability that a photon will have traveled a distance s without having interacted with the matter. The probability that a photon will travel a distance s without interaction and then interact in the next interval Δs is thus

$$q(s) \cdot n\sigma\Delta s. \tag{1.20}$$

Therefore the average length of a photon flight is simply

$$\langle s \rangle = \frac{\int s\exp(-n\sigma s)n\sigma ds}{\int \exp(-n\sigma s)n\sigma ds} = \frac{1}{n\sigma}, \tag{1.21}$$

and the quantity $1/n\sigma$ is referred to as the mean free path. Evidently the quantity $\tau = x/(n\sigma)^{-1} = n\sigma x$ is simply the number of mean free paths. Even though n is a function of x we may still introduce τ defined by

$$\tau = \int_0^x n\sigma ds \tag{1.22}$$

so that equations (1.18) and (1.19) then may be written

$$\frac{dI}{d\tau} = -I \; ; \quad I = I_o e^{-\tau} . \tag{1.23}$$

The dimensionless variable τ is called the <u>optical depth</u> or <u>optical thickness</u>.

Our aim is to derive equations which will enable us to solve for the specific intensity $I_\nu(x,\theta)$. So far we have seen that the radiation field decreases exponentially with distance along a given direction, but this neglects the emission of new photons which tend to replenish those absorbed and scattered. We define a <u>volume emission coefficient</u> $\varepsilon_\nu(x, \theta)$ as follows:

$$\Delta E = \varepsilon_\nu(x,\theta)\Delta V\Delta\Omega\Delta t\Delta\nu \tag{1.24}$$

is the energy emitted in volume $\Delta V = \Delta A\Delta s$ within solid angle $\Delta\Omega$ in time Δt and frequency interval $\Delta\nu$. Comparing this with the definition of the specific intensity, equation (1.4), we see that this emitted energy represents an increment in the specific intensity over a distance Δs of $\Delta I_\nu = \varepsilon_\nu \Delta s$. Hence, combining the expressions for both the increase and decrease in the beam we obtain the basic equation of radiative transfer

$$\frac{dI_\nu}{ds} = - I_\nu n\sigma + \varepsilon\nu . \qquad (1.25)$$

We have previously discussed the conceptual difference between <u>scattering</u> and <u>absorption</u>. From our discussion of scattering it is clear that there is a simple connection between the attenuation of the photon beam, and the redirection or "emission" of other scattered photons originally traveling in other directions. (The term "absorption" would normally be used here to describe the attenuation, but in this case we do not use it because we wish to distinguish scattering and absorption). The amount of energy removed from a beam of photons traveling in some particular direction θ' and confined to some solid angle $\Delta\Omega'$ is

$$\Delta E' = (\Delta I_\nu')\Delta A'\Delta\Omega'\Delta\nu\ \Delta t = (n\sigma_s I_\nu'\Delta s')\Delta A'\Delta\Omega'\Delta\nu\ \Delta t. \qquad (1.26)$$

This energy is then redistributed in direction (uniformly we will here assume) over 4π steradians so that the energy redirected into $\Delta\Omega$ is

$$\frac{\Delta E'\Delta\Omega}{4\pi} = n\sigma_s I_\nu'\ \Delta s'\Delta A'\Delta\Omega'\Delta\nu\Delta t\ \frac{\Delta\Omega}{4\pi} . \qquad (1.27)$$

Evidently the energy scattered into $\Delta\Omega$ from all initial directions is found by integrating equation (1.27) over $\Delta\Omega'$. Comparing this expression with the definition of the volume emission coefficient (1.24) and recalling the definition of the mean intensity, J_ν, equation (1.15), we see that

$$\varepsilon_{\nu sc} = n\ \sigma_s\ J_\nu. \qquad (1.28)$$

The treatment of the portion of the emission coefficient associated with a "pure absorption" process is based upon the supposition that the populations of the two quantum states involved in the absorption and emission of the photon are as they would be in thermodynamic equilibrium. In that case, as we shall discuss below, Kirchhoff's law holds, and the emission and absorption coefficients are simply related by

$$\varepsilon_{\nu,a} = n\ \sigma_a B_\nu(T), \qquad (1.29)$$

where B_ν is the Planck function. The transfer equation can then be written

$$\frac{dI_\nu}{ds} = -[n\sigma_s + n\sigma_a]I_\nu + n\sigma_s J_\nu + n\sigma_a B_\nu . \quad (1.30)$$

It should be stressed that the division of absorption coefficients and the corresponding emission coefficients into "scattering" and "pure absorption" is simply an approximation. It is not a bad approximation, for example, in the solar atmosphere in the vicinity of a strong resonance line (e.g. the sodium "D" lines, or Calcium ion "H" and "K" lines) which simply scatter, while a few photons at these wavelengths ionize the H^-ion, and this latter process and the corresponding emission process obey the relation (1.29) with fair accuracy. However, in computing the emission of, say, $H\alpha$ in a planetary nebulae, such a division is useless.

We have seen previously that the variable $d\tau = n(\sigma_s + \sigma_a)ds$ is a useful one to introduce. If we do this in equation (1.30) then we obtain

$$\frac{dI_\nu}{d\tau} = -I_\nu + S_\nu , \quad (1.31)$$

where

$$S_\nu = \frac{\varepsilon_{\nu,s} + \varepsilon_{\nu,a}}{n(\sigma_s + \sigma_a)} = \frac{n\sigma_s J_\nu + n\sigma_a B_\nu}{n\sigma_s + n\sigma_a} . \quad (1.32)$$

The quantity S_ν is called the <u>source function</u>, and its physical significance is that it represents the increment in the specific intensity due to the emission over a distance of one mean free path. Note that for pure absorption $S_\nu \equiv B_\nu$ and for pure scattering $S_\nu \equiv J_\nu$.

A formal solution to the equation of transfer can be written if we know $S_\nu(t)$, and I_ν at some point, $I_{\nu_0}(\tau_0) \equiv I_0$, since equation (1.31) is a first order, linear, inhomogeneous differential equation whose solution is

$$I(\tau) = I_0 \exp[-(\tau-\tau_0)] + \int_{\tau_0}^{\tau} S_\nu(\omega)\exp[-(\tau-\omega)]d\omega . \quad (1.33)$$

A simple interpretation of this equation is as follows: the contribution to the specific intensity $I_\nu(\tau)$ is made up of the known intensity I_o at some point τ_0 attenuated by the absorption between τ_0 and τ plus the additional radiation produced between τ_0 and τ, each interval between ω and $\omega + \Delta\omega$ producing a contribution $S_\nu(\omega) d\omega$ which is itself attenuated by the factor $\exp[-(\tau - \omega)]$.

To develop some feeling for the relation between the "emergent intensity" and the source function, let us consider the following simple cases, all with $I_o = 0$. If

a) S_ν = constant over some frequency range, and the slab has thickness τ^* and $\tau_0 = 0$. Then

$$I_\nu = S_\nu [1 - \exp(-\tau^*)], \tag{1.34}$$

$$\tau^* << 1, \; I_\nu \simeq S_\nu \tau^* = (\varepsilon_\nu / n\sigma)(n\sigma x) = \varepsilon_\nu x, \tag{1.34a}$$

$$\tau^* >> 1, \; I_\nu \simeq S_\nu . \tag{1.34b}$$

b) S is a linear function of τ, $t^* >> 1$ and $S = S_0 - S_1 \tau$ for $-\infty < t < 0$,

$$I_\nu = S_0 + S_1 = S_\nu (\tau = -1). \tag{1.35}$$

This latter result is quite important: It says that if, over a few optical depths near the surface of an optically thick medium ($\tau^* >> 1$) the source function varies in a sufficiently smooth manner that a linear approximation to the source function is fairly good, then the value of the emerging specific intensity is equal to the value of the source function at one optical depth below the surface. Note that thus far the optical depth is measured along the direction of the radiation under consideration.

1.3 "Microscopic" Formulation of the Transfer Equation

Evidently, determining the source function is the key to the solution of the transfer equation. In order to do this properly, we must examine in more detail the emission and absorption of radiation. That is, we must consider a "microscopic" formulation for the emission and absorption of radiation which considers the population of the atoms in various energy levels. We now consider some important types of matter-radiation interactions in "low-energy" astrophysics--that is, where the energy of the photon is not more than a few hundred electron volts.

Line Absorption and Emission. The strength of the transition from one discrete energy level is described in terms of the Einstein A and B coefficients, which involve the matrix elements of the electromagnetic field-atom interaction connecting the two states in question. The Einstein A coefficient, $A_{u\ell}$ is the probability per second of a spontaneous transition to the lower level. In the absence of other processes one has

$$\frac{dN_u}{dt} = - N_u \sum_{\ell<u} A_{u\ell} \, , \tag{1.36}$$

where N_u is the number density of particles in the sample in the upper state, u. Note that the N defined above refer to individual quantum states, not to a group of degenerate states; in the latter case, the statistical weights will be involved in equation (1.39). 1/A is the mean life-time of the upper state against spontaneous transitions to the lower state. For strong optical transitions $A \simeq 10^6$ to 10^8 sec^{-1}. The Einstein B coefficient for absorption may be defined such that the number of absorptions per second to higher levels is given by

$$\frac{dN_\ell}{dt} = -N_\ell \sum_{u>\ell} B_{\ell u} \int J_\nu \, \phi_\nu \, d\nu. \tag{1.37}$$

J_ν is the mean intensity, and ϕ_ν is the normalized line profile to be described below. Finally, in addition to spontaneous emission, the presence of photons will cause stimulated emission, involving another coefficient $B_{u\ell}$. The number of stimulated emissions to lower levels is

$$\frac{d}{dt} N_u = - N_u \sum_{\ell<u} \int J_\nu \phi_\nu d\nu \, . \tag{1.38}$$

The three coefficients are not independent, since they all involve the same matrix element. (In fact, if the B's are defined in terms of the occupation index, η rather than J_ν, then $A_{u\ell} = B_{u\ell} = B_{\ell u}$). In terms of the definition of B involving J,

$$A_{u\ell} = (2h\nu^3 / c^2)\, B_{u\ell} \;;\; B_{u\ell} = B_{\ell u} \,. \tag{1.39}$$

Equation (1.39) is a result following directly from quantum mechanics, but this result is usually derived by a thermodynamic argument involving the principle of detailed balance, which we shall illustrate below.

To see the connection between the Einstein B coefficients and the cross section we note that the rate at which energy is absorbed in an isotropic radiation field in unit volume is

$$\frac{dE}{dt} = 4\pi N \int \sigma(\nu) J(\nu)\, d\nu = h\nu_o N B_{\ell u} \int J_\nu \phi_\nu d\nu \;, \tag{1.40}$$

where ν_o is the line center and the variation in the energy $h\nu$ over the line has been ignored on the right hand side. Hence, by considering the special case in which J is constant across the line, we conclude that

$$\int \sigma(\nu) d\nu = \frac{h\nu_o B_{\ell u}}{4\pi} = \frac{\lambda_o^2 A_{u\ell}}{8\pi} \;. \tag{1.41}$$

Typically, $\sigma(\nu)$ is a sharply peaked function of width $\Delta\nu$ (with $\Delta\nu \sim 10^{10}$ sec^{-1} typically) so that, roughly, near the line center

$$\sigma(\nu) \simeq \frac{\lambda^2}{\Delta\nu} \frac{A_{u\ell}}{8\pi} \;. \tag{1.42}$$

Thus, for strong optical lines typical values of the cross section are in the range 10^{-12} to 10^{-13} cm^2.

A purely classical treatment leads to the result (see Wooley and Stibbs, pp. 91-94) that $\int \sigma(\nu) d\nu = \frac{\pi e^2}{m_e c} =$ .026 cm^2 sec^{-1}, and it is therefore also customary to define the exact quantum mechanical result for absorption in terms of the f-value:

$$\int \sigma(\nu) d\nu = \left(\frac{\pi e^2}{mc}\right) f. \tag{1.43}$$

Continuous Absorption. The strength of the continuous absorption process, or photoionization is generally given directly in terms of the cross-section $\sigma(\nu)$. If $\chi = h\nu_0$ is the ionization potential, then of course no photoionization can take place for $\nu < \nu_0$ and $\sigma(\nu) = 0$ in this frequency range. For most quantum states of most ions, $\sigma(\nu)$ can be described roughly by

$$\sigma(\nu) = \sigma(\nu_0)\left(\frac{\nu_0}{\nu}\right)^m \quad \text{for } \nu > \nu_0 \quad , \tag{1.44}$$

with m typically 2-3 (it is about 3 for hydrogen) and typical values of $\sigma(\nu_0)$ being 10^{-17} -10^{-18} cm^2 . Thus, the cross section for photoionization is usually several orders of magnitude smaller than the cross section near a strong line, although the values of $\int\sigma(\nu)d\nu$ are comparable. The rate of the inverse process, involving the capture of a free electron by an ion to some particular quantum state, (i.e. the number of such captures per cm^3/sec) is given by

$$N_i N_e \langle\sigma_r v\rangle = N_i N_e \int_0^\infty \sigma(E) v(E) f(E,T_e) dE \quad , \tag{1.45}$$

where $\langle\sigma_r v\rangle$ is the recombination rate, $\sigma(E)$ the recombination cross section for electrons of kinetic energy E and velocity $v(E)$, and $f(E, T_e)$ is the distribution function for electrons (generally assumed to be Maxwellian), and T_e is the electron temperature. The values of σ are generally quite small, typically 10^{-21} cm^2 . "Stimulated recombination" can also occur.

Free-Free Absorption and Emission. In the classical description of the process of free-free emission, an electron moving under the influence of a nearby ion accelerates and the accelerated charge radiates energy. The rate of radiation per cm^3 is thus proportional to the number of ion-electron encounters and hence proportional to $N_i N_e$. Quantum mechanically, the electron makes a transition from one continuum state to another of lower energy, emitting a photon. In the reverse process an electron in a low continuum energy state associated with an ion absorbs a photon, which also depends upon the number of ion-electron encounters. As mentioned previously, the concept of a "cross-section" is not especially appropriate in this case, and one simply uses the linear absorption coefficient:

$$\alpha = N_i N_e \int \psi(\nu,E) f(E,T) dE = N_i N_e \kappa(\nu,T) . \tag{1.46}$$

For ν in the optical frequency range: $\kappa(\nu,T)=3.7\times10^8 Z^2 T^{-\frac{1}{2}}\nu^{-3}$, Z being the charge on the ion. In the circumstances usually encountered in stellar atmospheres and gaseous nebulae, bound-free absorption is usually far more important than free-free absorption in the visible, but free-free absorption becomes very important in the far infrared and radio wavelengths.

<u>Electron Scattering.</u> The cross section for scattering by free electrons whose energy is small compared to the rest energy of the electron is

$$\sigma = 6.66 \times 10^{-25}\ \mathrm{cm}^2. \tag{1.47}$$

("Stimulated scattering" occurs, but is not generally an important consideration). Note the very small value of this cross section compared to that for line, or even bound-free absorption. Nevertheless in circumstances where the gas is very highly ionized and there are virtually no particles with bound states (as in the atmosphere of a very hot O star), electron scattering may be the dominant source of opacity.

<u>Collision cross sections.</u> Although not directly involving creation or destruction of photons, knowledge of the cross section for the excitation of an atom from one quantum state to a higher one is very important in astrophysics, especially in low-density situations (e.g. H II regions, planetary nebulae). The rate of excitation is expressed in the same form as that of equation (1.45), except that the lower limit on the integral is χ, the excitation energy, since the incoming electron must have this minimum energy to excite the transition. Typically, the cross-section for collisional excitation is about $10^{-16}\ \mathrm{m}^2$, even for cases where the radiative transition is very weak ("forbidden"). The inverse process, collisional de-excitation, involves an incoming electron colliding with an atom or ion in an excited state, leaving it in a state of lower energy, the energy difference appearing as excess energy of the incoming electron.

We now illustrate the principle of detailed balance referred to previously to calculate the relation between rates of excitation and de-excitation of atoms by electron collisions. To every one of the collision, absorption, or scattering processes we have enumerated there is an inverse--the process corresponding to the one in question if we simply reverse the direction of the flow of time. As previously indicated the formalism of quantum mechanics directly establishes the connection between the rate of any process and its inverse process (being formally traceable to the Hermitian character of the interaction matrix element), saving us, in effect half the trouble of computing all desired rates. However a somewhat more indirect argument is often used to establish this connection. The principle of detailed balance asserts that in thermodynamic equilibrium each process occurs at exactly the same rate as its inverse process (see, for example, Fowler, 1955, pp. 659-660).

To illustrate this, consider a collisional excitation cross-section and the corresponding inverse collisional de-excitation cross-section. Equating the rates for each energy increment, we have

$$N_e N_\ell \sigma(\ell \to u; E) V(E) f(E) = N_e N_u \sigma(u \to \ell; E-\chi) V(E-\chi) f(E-\chi), \quad (1.48)$$

where χ is the energy difference between the upper and lower state. In thermodynamic equilibrium we have (we assume Maxwellian statistics; strictly speaking the particles will obey Fermi-Dirac or Bose-Einstein statistics)

$$N_u/N_\ell = e^{-\chi/kT}; \quad f(E) = \frac{2}{\sqrt{\pi}} \frac{1}{(kT)^{3/2}} E^{1/2} e^{-E/kT}, \quad (1.48a,b)$$

from which we infer that

$$\sigma^{ex}_{\ell u}(E) = \frac{(E-\chi)}{E} \sigma^{ex}_{u\ell}(E - \chi). \quad (1.49)$$

In most low-energy astrophysical situations (1.48b) holds, though (1.48a) may not. In such a situation, equation (1.49) can be used to infer that the rates of collisional excitation and de-excitation are simply related by

$$\langle\sigma^{ex}_{\ell u} v\rangle = N_e \int_E^\infty \sigma^{ex}_{\ell u}(E)\, f(E,T_e) v(E) dE = e^{-\chi/kT_e} \langle\sigma^{ex}_{u\ell} v\rangle. \quad (1.50)$$

In connection with the application of detailed balance two points should be made:

a) relations such as (1.49) are valid regardless of whether either (1.48a) or (1.48b) obtains--that is, the assumption of thermal equilibrium was just a device to obtain the result, (1.49), which holds in any situation;

b) in using detailed balance to derive relations between emission and absorption processes, spontaneous and stimulated emissions should be considered lumped together as one process, since spontaneous emission is to be regarded as emission stimulated by the "zero-point" radiation field (see Leighton Sections 6-2, 6-3, 6-4).

Let us now consider the transfer equation between two bound states, and for simplicity assume the frequency profile of the emission and absorption coefficients near the line center to be the same. Let us further assume that these two states are the only quantum states. Then the absorption coefficient is (see equation 1.40) $B_{\ell u} h\nu_o \phi_\nu/4\pi$, the stimulated emission coefficient is $B_{u\ell} \cdot h\nu_o \phi_\nu/4\pi$ and the spontaneous emission coefficient is $A_{u\ell}$ $h\nu_o \phi_\nu/4\pi$ with ϕ_ν normalized to unity so that $\int \phi_\nu \, d\nu = 1$ Hence the transfer equation in terms of the Einstein A's and B's is making use of equation (1.39):

$$\frac{dI_\nu}{ds} = -\phi_\nu I_\nu h\nu \frac{B_\nu}{4\pi} N_\ell \left[1 - \frac{N_u}{N_\ell}\right] + \frac{2h\nu^3}{c^2} B_\nu \ . \tag{1.51}$$

To proceed with the solution of equation (1.51) it is evidently necessary to know how the atoms are distributed over the energy states N_u and N_ℓ. Unless there is a rather strong time dependence in the general environment, it is adequate to assume that statistical equilibrium exists: that is, that the population of any of the states is constant in time. The condition for this to be so in our simple case is

$$N_\ell (C_{\ell u} + B_{\ell u} \int J_\nu \phi_\nu d\nu) = N_u (C_{u\ell} + B_{u\ell} \int J_\nu \phi_\nu d\nu + A_{u\ell}), \tag{1.52}$$

where $C_{\ell u} = N_e \langle \sigma^{ex}_{\ell u} v \rangle$ and $C_{u\ell} = N_e \langle \sigma^{ex}_{u\ell} v \rangle$. The concept of statistical equilibrium should be carefully distinguished from that of detailed balance, which is a far more restrictive condition. Let us make one further final simplification in order to draw some conclusions from equations (1.51) and (1.52): that J_ν is about constant over the region where ϕ_ν is large, so that we write, using equations (1.39) and (1.52)

$$\frac{N_u}{N_\ell} = \frac{C_{u\ell}e^{-\chi/kT_e} + A_{u\ell}\ [J_\nu/(2h\nu^3/c^2)]}{C_{u\ell} + A_{u\ell}\ [1 + J_\nu/(2h\nu^3/c^2)]} \quad . \tag{1.53}$$

As before, we may introduce the optical depth and source function

$$d\tau_\nu = \phi_\nu BN_\ell \frac{h\nu_o}{4\pi}\ [1 - \frac{N_u}{N_\ell}]; \tag{1.54a}$$

$$S_\nu = \frac{2h\nu^3}{c^2}\ \frac{N_u/N_\ell}{[1-N_u/N_\ell]} \quad . \tag{1.54b}$$

Note the following points:

a) Assuming a Maxwellian distribution of electron velocities (so that equation 1.50 is valid) equation (1.53) says that, provided the collisional terms dominate the radiative terms, (which will generally be true if $C_{u\ell} >> A_{u\ell}$), then $N_u/N_\ell = e^{-\chi/kT}$.

b) If J is equal to the Planck function then $N_u/N_\ell = e^{-\chi/kT_e}$ regardless of the relative size of C and A.

c) If $N_u/N_\ell = e^{-\chi/kT_e}$, then S_ν is equal to the Planck function. The assumption that S_ν equals the Planck function is called the assumption of local thermodynamic equilibrium (LTE). In this case it is evidently not necessary to solve statistical equilibrium equations like equation (1.52), but simply to be given the temperature as a function of τ, in order to solve the transfer equation. All other situations are referred to as non-local thermodynamic equilibrium (NLTE). Evidently a sufficient condition for LTE to be a good approximation is for collisional rates to dominate over radiative ones: in general, this means that L.T.E. is a fairly good approximation for sufficiently high densities.

d) Note that the simulated emission factor $[1-N_u/N_\ell]$ in equation (1.51) reduces the effective value of the absorption coefficient. In terms of the excitation temperature (defined, in general, by $N_u/N_\ell = e^{-\chi/kT_{ex}}$), which is equal to the kinetic temperature of the electrons for LTE, $[1 - N_u/N_\ell] = 1 - e^{-\chi/kT_{ex}}$, which for $\chi/kT_{ex} << 1$ reduces to χ/kT_{ex}, and hence in these situations is a very important correction. If $N_u > N_\ell$ then the absorption is negative and we get exponential growth, (amplification) instead of attenuation, which occurs in "astrophysical masers".

e) We have glossed over a number of subtle but important points in the preceding discussion having to do mainly with the shape, or profile, of the emission and absorption coefficients. The absorption profile ϕ_ν is fairly straightforward (though complicated), and we will discuss it further below. The spontaneous emission profile is more involved: its shape depends, roughly speaking, upon the 'history' of the particular photon being emitted: If it is simply scattered, then the emission profile takes the form

$$\psi(\nu) \simeq \int R(\nu,\nu')J_{\nu'}d\nu' , \tag{1.55}$$

where R is the so-called redistribution function (cf. Mihalas 1970, Chapter 10). On the other hand, for those photons arriving in the upper state as the result of collisions, the emission profile is the same as the absorption profile. (It is also usually stated and assumed that the stimulated emission profile is the same as the absorption profile, but this need not be the case). Thus, the correct formulations of (1.51) and (1.52) must take into account this complication in the emission profile.

The treatment of the full set of NLTE transfer and statistical equilibrium equations for many-level atoms is a formidable problem, and much current research in stellar atmospheres is devoted to dealing with this problem (see Mihalas 1970 and Jefferies 1968).

1.4 A Simple Solution to the Transfer Equation

We have so far developed some of the formalism of radiative transfer and now wish to apply it to a simplified atmosphere in <u>radiative equilibrium</u>. Let us take as the form of the transfer equation, equation (1.30):

$$\frac{dI_\nu}{ds} = - [n\sigma_s + n\sigma_a] I_\nu + n\sigma_s J_\nu + n\sigma_a B_\nu \ . \tag{1.56}$$

In this equation, s is the distance along the particular ray in question. If the atmosphere has plane symmetry, as we shall assume, it is convenient to introduce the optical depth measured normal to the layers. (Very often one finds τ defined with a minus sign, so that τ increases as one moves down into the atmosphere):

$$d\tau_\nu = (n\sigma_s + n\sigma_a) dx = (n\sigma_s + n\sigma_a) \cos\theta \ ds. \tag{1.57}$$

Hence

$$\mu \frac{dI_\nu}{dx} = - [n\sigma_s + n\sigma_a] I_\nu + n\sigma_s J_\nu + n\sigma_a B_\nu \ , \tag{1.58}$$

or

$$\mu \frac{dI_\nu}{d\tau_\nu} = - I_\nu + \lambda_\nu B_\nu + (1-\lambda) J_\nu \ , \tag{1.59}$$

where

$$\lambda_\nu = \frac{n\sigma_a}{n\sigma_a + n\sigma_s} \ . \tag{1.60}$$

Let us now integrate equation (1.58) over μ. The result is

$$\frac{dH_\nu(x)}{dx} = - n\sigma\lambda_\nu (J_\nu - B_\nu) = - n\sigma_{ab} (J_\nu - B_\nu). \tag{1.61}$$

Similarly, if we first multiply equation (1.58) by μ and then integrate over μ we obtain

$$\frac{dK_\nu(x)}{dx} = - n\sigma(\nu) H_\nu \ . \tag{1.62}$$

Next, let us integrate (1.61) and (1.62) over all frequencies to get:

$$\frac{dH}{dx} = - n \int \sigma_a(\nu) J_\nu d\nu + n \int \sigma_a(\nu) B_\nu(T) d\nu , \tag{1.63}$$

$$- \frac{dK}{dx} = n \int \sigma(\nu) H_\nu d\nu . \tag{1.64}$$

Note that the left hand side of equation (1.63) is just the divergence of the radiative flux. In a star there may be sources of heat (e.g. nuclear energy) and/or other means of transporting energy besides radiation. If such is not the case however, then conservation of energy demands that the divergence of the flux be zero. Layers of a star in which the divergence of the radiative flux is zero are said to be in radiative equilibrium. The condition for radiative equilibrium is thus

$$\int \sigma_a(\nu) J_\nu d\nu = \int \sigma_a(\nu) B_\nu(T) d\nu . \tag{1.65}$$

In connection with equation (1.64), if we multiply by $4\pi/c$ we obtain

$$- \frac{dP}{dx} = \frac{n}{c} \int \sigma(\nu) F(\nu) d\nu . \tag{1.66}$$

The divergence of the pressure is the force density associated with the radiation, from which it follows that the acceleration associated with radiation incident on a particle of cross section σ and mass m is just

$$A_{rad} = \int \sigma(\nu) F_\nu d\nu / mc . \tag{1.67}$$

An atmosphere in which σ_s and σ_a are independent of ν is called a grey atmosphere. It follows from (1.65) that for a grey atmosphere in radiative equilibrium $J = \int J_\nu d\nu = \int B_\nu d\nu = B = \sigma_B T^4/\pi$, where σ_B is the Stefan-Boltzmann constant. If, for this case we integrate equation (1.58) over frequency then we obtain simply

$$\mu\frac{dI}{d\tau} = - I + J = - I + 1/2 \int_{-1}^{+1} I(\mu';\tau)d\mu \quad . \qquad (1.68)$$

This classic form of the transfer equation is sometimes known as "Milne's equation". In the usual case of a semi-infinite atmosphere with no incident radiation the boundary conditions are that $I(\mu, \tau \equiv 0) = 0$ for $\mu<0$, and that $J(\tau)$ be "well behaved" as $\tau \to -\infty$. As formulated the problem is an integrodifferential equation for I; however it can readily be converted to an integral equation for J. An exact analytic solution of equation (1.68) has been obtained (see Wooley & Stibbs, 1953 Chapter 3).

A simple approximate solution to many transfer equations can be obtained, based upon the following consideration. As we go deeper and deeper into the atmosphere, we expect the influence of the surface on the directionality of the photon field to be less and less. In fact, except very near the surface, we expect I_ν to be more-or-less isotropic, in which case $K_\nu = 1/3\, J_\nu$. Of course, it cannot be anywhere perfectly isotropic if there is to be a flow of energy; however provided we can approximate $I_\nu(\mu, \tau)$ by $I_0 + I_1\mu$ then, as we see from applying equation (1.15), it is still true that $K_\nu = 1/3 J_\nu$. This is the Eddington approximation. Equations (1.61) and (1.62) then become

$$\frac{dH_\nu}{d\tau_\nu} = - \lambda_\nu(J_\nu - B_\nu), \qquad (1.69)$$

$$\frac{dJ_\nu}{d\tau_\nu} = - 3H_\nu(T_\nu) \quad . \qquad (1.70)$$

For the particular case of Milne's equation (equation 1.68) we have $\frac{dH}{d\tau} = 0$ so that $H = H_c$ and $J = - 3H_c\tau + C$.

The constant H_c is the constant flux which must be determined from other considerations (e.g. by considering the energy generation in the interior of the star) and the constant C must be determined by the outer boundary condition. The simplest approach to the outer boundary condition is to assume that I is isotropic over $0 < \mu < 1$ (and is of course zero $-1 < \mu < 0$), in which case $H(\tau = 0) = \frac{1}{2}J(\tau = 0)$, so that $C = 2H_c$.
Thus,

$$J(\tau) = H_c[-3\tau + 2], \quad -\infty<\tau<0, \tag{1.71}$$

which also specifies how T must vary with τ. A more systematic analysis by Krook (1966) shows that it is more consistent to use $J = \sqrt{3}H$ at $\tau = 0$ as the boundary condition. We shall approach the Eddington Approximation from a slightly different point of view in considering transfer of radiation in the stellar interior; we remark here that the Eddington approximation is really equivalent to a diffusion approximation in kinetic theory.

1.5 Model Stellar Atmospheres

We now outline the ideas behind the construction of a model atmosphere. We begin the construction by making the following assumptions:

(1) The atmospheric layers are in hydrostatic equilibrium.
(2) The atmosphere is in radiative equilibrium.
(3) Divide the absorption (absorption & scattering coefficients) into two components: continuous absorption and line absorption. As far as the overall structure of the atmosphere is concerned, it is frequently the case that although the line absorption coefficients are very strong, the lines nevertheless interact with only a small amount of the total flux passing through the atmosphere, because of the sharpness of the lines. Hence they exert only a modifying influence on the overall structure of the atmosphere and are often ignored in a

first approximation.

(4) The continuous opacity (κ_ν) can be divided into a scattering and absorption component $(\kappa_{\nu,s}$ and $\kappa_{\nu,a})$.

(5) The atmosphere is plane parallel.

From the first and fifth assumption we obtain (see Chapter 2) for a discussion of hydrostatic equilibrium):

$$\frac{dP}{dx} = - \rho g, \tag{1.72}$$

where P is the pressure, ρ the gas density and g the acceleration of gravity. In addition we have the relation

$$\frac{d\tau_\nu}{dx} = - n\sigma_\nu = - \rho\kappa_\nu, \tag{1.73}$$

where we have now measured τ_ν inward, and have defined a new quantity, κ_ν, the mass absorption coefficient. The mass absorption coefficient can be thought of as the total cross section for one gram of matter, as opposed to one particle. From equations (1.72) and (1.73) we obtain

$$\frac{dP}{d\tau_\nu} = \frac{g}{\kappa_\nu} = \frac{g}{\kappa_{\nu,s}+\kappa_{\nu,a}} . \tag{1.74}$$

We also have from assumption (2) the condition of radiative equilibrium (1.69) which we write as

$$\int\kappa_{\nu,a}B_\nu(T)\,d\nu - \int \kappa_{\nu,a}J_\nu\,d\nu = 0. \tag{1.75}$$

In addition, of course, we have the transfer equation itself

$$\mu\frac{dI_\nu(\tau_\nu,\mu)}{d\tau_\nu} = I_\nu - \lambda_\nu B_\nu(\tau_\nu)-(1-\lambda_\nu)1/2\int_{-1}^{+1}I_\nu(\mu,\tau_\nu)\,d\mu \ , \tag{1.76}$$

with

$$I(\mu,\tau=0) \equiv 0, \; -1<\mu<0 \quad .$$

We must also be able to compute a table of absorption coefficients giving $\kappa_{\nu,s}$ and $\kappa_{\nu,a}$ as a function of ν, ρ, T and CC (or, through an equation of state, (see Chapter 2) connecting P, ρ, T, of P, T, and CC) where CC stands for chemical composition. The pressure and temperature enter the determination of κ_ν because, through the assumption of LTE implicit in our division of κ_ν into $\kappa_{\nu,a}$ and $\kappa_{\nu,s}$, P and T determine how the population of each element is distributed over the various ionization stages, and quantum states within each ion, and we are assuming LTE insofar as continuous absorption and emission are concerned. Thus we assume that we can make two tables, which we write symbolically as

$$\kappa_a = \kappa_a(P,T,CC), \tag{1.77a}$$

$$\kappa_s = \kappa_s(P,T,CC). \tag{1.77b}$$

Note that while we have used τ_ν as "the" independent variable in equations (1.74) and (1.76), a given value of τ_ν (say $\tau_\nu = 1$) will in general refer to very different values of x, depending upon which ν we choose. Note that equation (1.75) must hold at every given value of x. It is convenient to think of some reference ν, say, ν_o and to consider τ_{ν_o} as the main independent variable; then we regard $T(\tau_{\nu_o})$ and $P(\tau_{\nu_o})$ as functions to be solved for, and can also write

$$\frac{d\tau_\nu(\tau_{\nu_o})}{d\tau_{\nu_o}} = \frac{\kappa[\nu,\, P(\tau_{\nu_o}),\, T(\tau_{\nu_o}),\, CC]}{\kappa[\nu_o,\, P(\tau_{\nu_o}),\, T(\tau_{\nu_o}),\, CC]} \quad . \tag{1.78}$$

Note finally that the radiative equilibrium condition, equation (1.75) only insures that the integrated flux $F = \int F_\nu \, d\nu$ is constant, but says nothing about what that constant is. Thus, to obtain a unique model atmosphere we must choose some F. It is more usual instead, however, to specify the effective temperature, related to F through

$$F = \sigma_B T_{eff}^4 , \tag{1.79}$$

where the effective temperature, is that temperature which a black body would have if it radiated a total flux F. Examination of equations (1.74) through (1.78) shows that we must also specify g and CC. If we specify T_{eff} , g, and CC then we may uniquely calculate a model atmosphere. This statement is the stellar atmosphere counterpart of the famous Russell-Vogt "theorem" of stellar structure discussed in Chapter 2.

The following (highly schematic!) steps are the ones taken to construct a model:

(a) Choose T_{eff} ($\Rightarrow$ F), g, CC.

(b) Construct opacity tables: $\kappa_a(\nu, P, T, CC)$; $\kappa_s(\nu, P, T, CC)$.

(c) Make an initial guess at the function $T(\tau_{\nu_0})$, using perhaps the $T(\tau_{\nu_0})$ obtained for the grey atmosphere.

(d) Integrate equation (1.74) through the atmosphere to obtain $P(\tau_{\nu_0})$, starting with the boundary condition that P = 0 when $\tau_{\nu_0} = 0$.

(e) Similarly integrate equation (1.78) to obtain $\tau_\nu(\tau_{\nu_0})$ at many different ν .

(f) Knowing $T(\tau_{\nu_0})$ and $\tau_\nu(\tau_{\nu_0})$ means that $T(\tau_\nu)$ and hence $B(\tau_\nu)$ is known: Then solve the integro-differential equation (1.76) to obtain $I_\nu(\mu, \tau_\nu)$ or $I_\nu(\mu, \tau_{\nu_0})$.

(g) Knowing $I_\nu(\mu, \tau_{\nu_0})$ we compute $J_\nu(\tau_{\nu_0})$ and then check to see how closely equation (1.75) is obeyed. Noting the (non-zero) value of the left hand side of equation (1.75) as a function of τ_0: $\Delta F(\tau_{\nu_0})$, make an improved choice of $T(\tau_{\nu_0})$ designed to make $\Delta F(\tau_{\nu_0})$ be zero everywhere.

(h) Return to (d) and repeat steps (d)-(g) until you are satisfied with how closely equation (1.75) is satisfied.

We have of course omitted here the many tricky, involved, and fascinating mathematical devices actually used in solving equation (1.76) in step (f), and in the "temperature correction procedure", step g. See the textbook by Mihalas (1970) for details.

1.6 Characteristics of Stellar Continuum Radiation

As a very rough zero'th approximation, we might expect the spectral energy distribution from the stellar continuum to resemble a black body of T_{eff}, and indeed this is true. This is not strictly true even for a grey body however, since even for a grey body in LTE, the emergent radiation is a weighted superposition of radiation emitted from gas with a range of temperatures.

More important departures from a black body arise in stars where the absorption coefficient varies significantly with wavelength. In A stars, for example, the so-called "Balmer Discontinuity" is a very prominent feature in the stellar continuum; the emitted flux for wavelengths a little less than 3650 Å is substantially less than the flux with wavelengths a little greater than 3650 Å. The reason can be traced to the result equation (1.35) which states that for a linearly varying source function ($S_\nu = S_0 + S_1\tau_\nu$) the emergent specific intensity I_ν $(o;\mu)$ has a value equal to S_ν $(t_\nu = 1)$ where t_ν is the optical depth measured along the ray in question and τ_ν the optical depth measured normal to the surface. We evidently have, since $\tau_\nu = \mu t_\nu$,

$$I_\nu(o,\mu) = S_\nu(t_\nu = 1) = S_0 + S_1\mu \ . \tag{1.80}$$

Therefore the emergent flux has the value

$$H_\nu = \frac{1}{4}[S_0 + \frac{2}{3}S_1] = \frac{1}{4}S_\nu(\tau_\nu = \frac{2}{3}), \tag{1.81a}$$

$$F_\nu = \pi S_\nu(\tau_\nu = \frac{2}{3}) \ . \tag{1.81b}$$

This result is sometimes known as the Eddington-Barbier relation.

Consider the case of the A star. At $\lambda_1 \simeq 3600$ Å, say, the opacity is rather high due to photoionization from the n = 2 level of hydrogen whereas as $\lambda_2 = 3700$ Å, (where photons have insufficient energy to cause photoionization from n = 2) the opacity is much less and is due to photoionization from n = 3; that is the cross section $\sigma_3(\lambda_2)$ is much less than $\sigma_2(\lambda_1)$. Since $\tau_\nu = \int m\sigma(\nu)dx \simeq \overline{n\sigma(\nu)}x$, the value of x at which $\tau_{\nu 1} = 2/3$ is much less (i.e. much closer to the top of the atmosphere) than the value of x at which $\tau_{\nu_2} = 2/3$. It thus follows that since the temperature is lower near the top of the atmosphere than in the deeper layers, the value of

$$S_{\nu_1}(\tau_{\lambda_1} = 2/3) < S_{\nu_2}(\tau_{\lambda_2} = 2/3) \text{ so that } F_{\lambda_1} < F_{\lambda_2} .$$

Essentially the same principle is involved in the principle of limb darkening. Referring to equation (1.80) we see that radiation emerging at an angle θ ($\cos\theta = \mu$) has a value equal to S_ν ($\tau_\nu = \mu$) . Hence radiation with μ near zero will be less intense than that emerging nearly normally ($\mu \simeq 1$) ($\theta \simeq 1$) using the same line of argument followed in discussing the Balmer discontinuity This effect is easily observed in the sun as we look from the center towards the edge or "limb" of the sun's disc.

To close this discussion of stellar continuum radiation we enumerate the principal sources of continuous opacity as a function of spectral class.

In the hottest O stars, the material is so hot that it is very highly ionized and electron scattering contributes strongly, despite its small cross section. Photoionization from singly ionized helium contributes significantly, and hydrogen photoionization makes a modest contribution, in spite of being highly ionized, due to its great abundance. As one moves down the main sequence, the photoionization from neutral helium becomes important, while electron scattering declines in importance. When the A stars are reached photoionization from hydrogen is by far the dominant opacity source. In the middle F stars through the K's, the dominant source of opacity is photoionization of the negative hydrogen ion, H^- whose sole bound state is a proton with two electrons orbiting it. In the very cool M stars, absorption processes involving molecules become important. It should be understood that the term "dominant source of opacity" refers to the main opacity source at wavelengths in the region where the source function is near its maximum; the value of the opacity at 500 Å in a layer of a star where the temperature of the star is 4000° is irrelevant to the radiative transport of energy through that layer.

Finally it may be mentioned that it is possible to define various averages of $K(\nu)$ over ν in an attempt to convert the problem of the non-grey atmosphere to the much simpler problem of the grey atmosphere. Except perhaps as starting points for the iteration procedure described above, these "mean absorption coefficients" are little used in modern model atmosphere construction. However, one of these "means" is very much used in stellar interiors, as described in Chapter 2.

1.7 Formation of Spectral Lines

In Section 1.3 we mentioned that the integrated value of the line absorption coefficient is, as we have seen, related to the f-value: $\int\sigma(\nu)d\nu = (\frac{\pi e^2}{mc})f$. Now, however, we wish to inquire about the shape of $\sigma(\nu)$, since that will affect the line profile.

It is convenient to think of $\sigma(\nu)$ as being influenced by three considerations:

(a) the intrinsic shape of $\sigma(\nu)$ associated with a single, isolated atom,

(b) the modification in the above $\sigma(\nu)$ due to perturbations associated with nearby particles, and

(c) the effect of the motion of the particles via the Doppler effect.

Regarding (a) both classical theory and the more precise quantum mechanical treatment yield the Lorentz profile (see Wooley and Stibbs 1953, Sections 2.1 and 4.3)

$$\sigma(\nu) = (\frac{\pi e^2}{mc})f\left\{\frac{8}{\pi}\frac{1}{(\nu-\nu_o)^2+\delta^2}\right\}, \tag{1.82}$$

where $\delta = \Gamma/4\pi$. Γ is the "natural damping coefficient", equal to the sum of the reciprocal of the mean lifetime Δt_L in the two quantum states. Crudely, the "half-half" width of the line is: $\Delta\nu_{1/2} = \delta = \Gamma/4\pi$; thus the energy width $\Delta E \simeq h\Delta\nu_{1/2} \simeq h\frac{\Gamma}{4\pi} \simeq \frac{h}{\Delta t_L}$, a result whose general form might have been anticipated on the basis of the uncertainty principle.

Regarding (b), an exact treatment is very complex, and the answer depends upon the particular transition and the kinds of perturbing particles. The resulting profile can generally be approximately expressed in the same form as (1.82) except that Γ consists of the sum of the reciprocal mean natural lifetime and mean lifetime against collisions. Since Γ is increased by these nearby particles, the profile is broader and the effect is variously described as "pressure broadening", "collisional broadening", or "Stark broadening" depending upon the particular transition and broadening agent.

Regarding (c) we consider a particular atom whose line center is at ν_o as seen in its rest frame. Consider a photon being emitted in our direction whose frequency we would observe to be $\nu = \nu_o + \Delta\nu$, where $\Delta\nu / \nu_o << 1$. Let this atom have a velocity component towards us of v_R ($v_R > 0$ if the atom is coming towards us, $v_R < 0$ if the atom is moving away from us). Then an observer in the atom's frame will observe this same photon to have a frequency

$$\tilde{\nu} = \nu - \nu_o \frac{v_R}{c} = \nu_o + \Delta\nu' ; \; \Delta\nu' = \Delta\nu - \nu_o \frac{v_R}{c} . \qquad (1.83)$$

Thus, the absorption coefficient that we would observe this particular particle to have would be given by

$$\sigma(\Delta\nu, v_R) = (\frac{\pi e^2}{mc}) f \; \{ \frac{\delta}{\pi} \frac{1}{(\Delta\nu - \nu_o \frac{v_R}{c})^2 + \delta^2} \} . \qquad (1.84)$$

Note from the sign that if v_R is positive (e.g. an outflowing shell of matter from a star) then the maximum absorption occurs when $\Delta\nu = \nu_o v_R / c > 0$, so that the photons blueward of what we consider the actual line center are the ones absorbed.

This is the result for a single particle: if we are dealing with a whole swarm of particles in a tiny volume of space then we can find an effective single particular absorption coefficient representing that volume of space by integrating equation (1.84) times the probability that a particle will have radial velocity in the range $v_R, v_R + dv_R$ over all possible radial velocities. If we assume that this motion is nothing but the thermal motion of the atoms, then

$$\sigma(\Delta\nu) = \frac{\pi e^2}{mc} \; f \; \frac{\delta}{\pi} \left(\frac{M}{2\pi kT} \right)^{1/2} \int_{-\infty}^{+\infty} \frac{\exp[-Mv_R^2/2kT]}{(\Delta\nu - \nu_o v_{R/c})^2 + \delta^2} dv_R \; . \qquad (1.85)$$

It is customary to measure frequency shifts in units of the Doppler width, $\Delta\nu_o = v_D \nu_o/c$: $x = \Delta\nu/\Delta\nu_D$ where $v_D = \sqrt{2kT/M}$ is the thermal Doppler velocity and M is the mass of the particle. Then

$$\sigma(x;a) = \left[\frac{\pi e^2}{mc}\frac{f}{\Delta\nu_D}\right]\frac{1}{\sqrt{\pi}}\left\{\frac{a}{\pi}\int_{-\infty}^{+\infty}\frac{e^{-y^2}dy}{(x-y)^2+a^2}\right\}, \tag{1.86}$$

where $a \equiv \delta/\Delta\nu_D$ is generally referred to as the damping parameter. Note that for $x = 0$, if we let $a \rightarrow 0$, the term $\{\ \} \rightarrow 1$. Thus

$$\frac{\sigma(x;a)}{\sigma_o} = \frac{a}{\pi}\int_{-\infty}^{+\infty}\frac{e^{-y^2}dy}{(x-y)^2+a^2} = V(x,a)\ , \tag{1.87}$$

with

$$\sigma_o \equiv \sigma(x=o,\ a=o) = \left(\frac{\pi e^2}{mc}\right)\frac{f}{\Delta\nu_D}\ \frac{1}{\sqrt{\pi}}\ . \tag{1.88}$$

The profile given by equation (1.86) is usually referred to as the Voigt profile; the integral $V(x, a)$ cannot be performed, but expansions in powers of a, analytic approximations, and extensive numerical tables exist. For our purposes the following observations are adequate for $a \ll 1$, which is generally the case in most astrophysical problems. The denominator in the integral (1.87) passes through a sharp minimum when $x = y$; provided a is not large, the main contribution to the integral will thus come near $x \simeq y$, and we can then treat $\exp(-y^2)$ as having the constant value $\exp(-x^2)$, perform the remaining integral with the result

$$\sigma(x,a) = \sigma_o e^{-x^2}\quad (x\ \text{"small"}). \tag{1.89}$$

For $x \gg 1$, on the other hand, the exponential is so small by the time $x \simeq y$ that the main contribution comes from $y \simeq 0$, in which case we may ignore a, put $(x - y)^2 \simeq x^2$ and thus obtain

$$\sigma(x,a) = \frac{\sigma_o}{\pi}\frac{a\sqrt{\pi}}{x^2} = (\frac{\pi e^2}{mc}) f \frac{\delta}{\pi} \frac{1}{(\Delta\nu)^2} \quad (x \text{ "large"}). \tag{1.90}$$

Generally, equation (1.89) is a fairly good approximation for $x \lesssim 2.5$, and equation (1.90) a good approximation for $x \gtrsim 3.5$, the transition region being rather sharp. The important thing to note is that in the Doppler core, $x \lesssim 3$ the profile drops off rapidly as e^{-x^2}, but then beyond $x \gtrsim 3$ there is a much gentler fall off, going as $1/x^2$.

To complete the picture of line formation we examine line profiles. Let us consider a semi-infinite atmosphere and the transfer equations (1.69) and (1.70) applied to line radiation and with $B = B_o + B_1\tau$ for both line and continuum in the vicinity of the line. The variations with frequency of B in the vicinity of the line can be ignored. Implicit in these equations is the assumption that the redistribution function (equation 1.55) is a delta-function, that is, that the scattering is "coherent", as well as the usual assumption of the validity of separations of κ into "scattering" and "absorption." Note that this special use of the word coherent--no change in frequency in a scattering process--is rather different from the usual meaning of the term.

The solution, with τ increasing inward is, using the Eddington approximation,

$$J = B_o + B_1\tau + \frac{[B_1/\sqrt{3}-B_o]}{1 + \sqrt{\lambda}} \exp(-\sqrt{3}\lambda\tau) \tag{1.91}$$

where we have used Krook's boundary condition (Krook 1955, page 24). What we are especially interested in now is the emergent flux

$$H(\tau=0) = \frac{1}{\sqrt{3}} \{ B_o + [B_1/\sqrt{3}-B_o]/[1+\sqrt{\lambda}] \} \tag{1.92}$$

Let us consider a spectral region just slightly removed from the region of the line, and let us assume for simplicity that $B_1 = \sqrt{3}B_o$ (for more realistic values of B_o/B_1 see

Swihart and Brown 1967). We have from equation (1.92)

$$H_o^c = B_o / \sqrt{3} = B_1^c /3. \tag{1.93}$$

where H_o^c is the flux and B_1^c is part of the source function outside the lines. Let us further assume that

$$\kappa = \kappa_a^L + \kappa_{sc}^L + \kappa_a^c , \tag{1.94}$$

where κ_a^c is the continuous absorption coefficient, essentially constant over the spectral region considered, and both κ_{sc}^L and κ_a^L are non-zero only in the line. Now the actual value of B at a given physical depth must be the same, hence

$$B_1^L \tau^L = B_1^c \tau^c , \tag{1.95}$$

so that

$$B_1^L = 3H_L^c \tau^c / \tau^L = \frac{3H_o}{1+\eta} , \tag{1.96}$$

where

$$\eta = \frac{\kappa_a^L + \kappa_{sc}^L}{\kappa_a^c} \tag{1.97}$$

is assumed constant.

Finally, we want to compute the ratio of the emerging flux in the line to that in the adjacent continuum

$$\frac{H_o^L}{H_o^c} = r = 1 + (\frac{1}{H\eta} - 1)/(1 + \sqrt{\lambda}) \tag{1.98}$$

where, in terms of (1.94)

$$\lambda = \frac{\kappa_a^L + \kappa_a^c}{\kappa_a^L + \kappa_a^c + \kappa_{sc}^L} . \tag{1.99}$$

Consider now the following special limiting cases:

(a) Pure absorption, $\lambda = 1$

(1) Weak line, $\eta \ll 1$

$$r \simeq 1 - \frac{\eta}{2} . \tag{1.100a}$$

(2) Strong line, $\eta >> 1$

$$r \simeq 1/2 . \tag{1.100b}$$

(b) Pure scattering in the line, $\lambda^{-1} = 1 + \eta$

(1) Weak line, $\eta << 1$

$$r \simeq 1 - \frac{\eta}{2} . \tag{1.100c}$$

(2) Strong line, $\eta >> 1$

$$r \simeq \frac{1}{\sqrt{\eta}} . \tag{1.100d}$$

Keep in mind that η will be a strong function of distance from the center of the line. The results for pure absorption follow our rough expectations following from the discussion of the Balmer discontinuity: even for very strong lines, the flux is not very much less than that of the continuum because we "see" radiation emitted from the top of the atmosphere; we expect something like this result on the basis of the Eddington-Barbier relation. The result for a strong line which scatters is a different matter: the line center can have nearly zero intensity, and we now consider an interesting way of looking at this.

It is possible to view radiative transfer from the point of view of the random walk of a photon, and this is an instructive way to view the result we have just obtained. Roughly speaking, we know that most of the radiation we see in the continuum is actually emitted in layers with $\tau_{cont} \lesssim 1$. From what layers are the line photons emitted that we actually see? The theory of random walk tells us that if, on the average a particle moves a distance ℓ between interactions it will, after N scatterings be, not a distance ℓN from its starting point but on the average a distance $\ell\sqrt{N}$. (Nevertheless, if we could straighten out its total path, it would have actually traveled a distance ℓN.) Given that our line photon suffered an interaction with either an absorber or scatterer, for the case (1.100d) we must assume that the probability is roughly $1/\eta$ that the photon will be absorbed --i.e., destroyed--rather than simply scattered. Therefore, the only emitted photons which we will actually see come

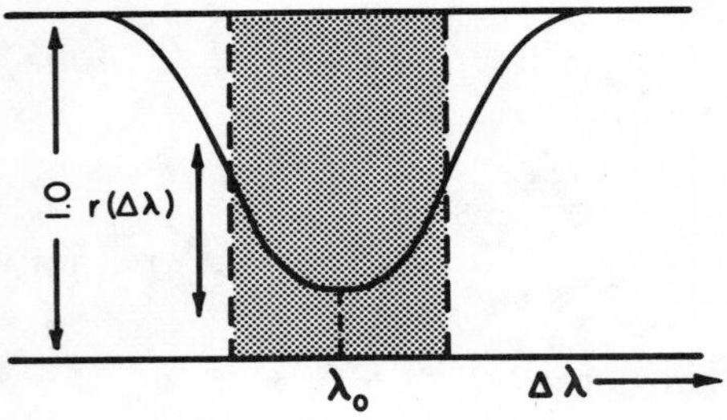

Fig. 1.5 Normalized line profile

from depths such that $N \lesssim \eta$. The distance ℓ is a length which is roughly $1/\eta$ in terms of the continuum optical depth. Those photons in the line therefore, which are to have an appreciable chance of being emitted and work their way out of the atmosphere after many scatterings before destruction, must be created at a continuum optical depth $\tau^c \lesssim \ell\sqrt{N} = \frac{1}{\eta}\sqrt{\eta} = \frac{1}{\sqrt{\eta}}$, so that the total flux in the line relative to the continuum is cut down by this amount. (This result is significantly altered when non-coherent scattering is considered).

1.8 The Curve of Growth

As our last topic in this Chapter we explore the concept of the curve of growth. This is a simple but powerful tool for exploring abundances of the elements in both interstellar matter and stellar atmospheres, a subject of obvious interest in connection with theories of nucleosynthesis (see Chapter 2).

Consider an absorption line whose continuum intensity in the vicinity of the line we agree to normalize to 1.0. Let the actual intensity at some distance $\Delta\lambda$ from the line center be r, so that the depression from the continuum is 1 - r. We define the equivalent width, W, by (cf. Fig. 1.5)

$$W = \int_{-\infty}^{+\infty} [1-r(\Delta\lambda)]d(\Delta\lambda), \tag{1.101}$$

W is evidently the area (In Å, say, since the height is normalized to 1) contained between the continuum and the line profile. Alternatively one can think of it as being the width of a completely black (r = 0) line having a rectangular profile, whose area is equivalent to the actual line.

The curve of growth is the relation between the equivalent width and, in the simple case of an absorption tube, the optical thickness at the line center. In a stellar atmosphere, this latter quantity makes no sense, but something equivalent (e.g. η , equation 1.97) can be used instead. Let us consider for simplicity, the very simplest case of a long thin "absorption tube" of length ℓ (like an interstellar cloud) which scatters essentially collimated light from a very distant source. In that case the source function is essentially zero and $r(\Delta\lambda)$ is simply given by

$$r(\Delta\lambda) = e^{-\tau(\Delta\lambda)}. \qquad (1.102)$$

Making use of equations (1. 87), (1.88), (1.101) and (1.102) and the definition of τ we have

$$W = \int_{-\infty}^{+\infty} \left\{ 1 - \exp[-n\ell\frac{\pi e^2}{mc} f \frac{1}{\Delta\nu_D} \frac{1}{\sqrt{\pi}} \cdot V(x,a)] \right\} d(\Delta\lambda) . \qquad (1.103)$$

Changing variables from $\Delta\lambda$ to X, and noting that $\Delta\nu_D = \frac{v_D \nu_o}{c} = \frac{v_D}{\lambda_o}$,

$$W(\tau_o, a) = \lambda_o \frac{v_D}{c} \int_{-\infty}^{+\infty} \left\{ 1\text{-}\exp[-\tau_o V(x,a)] \right\} dx , \qquad (1.104)$$

where

$$\tau_o = \frac{1}{v_D} \frac{\ell}{\sqrt{\pi}} \left(\frac{\pi e^2}{mc} \right) fn\lambda_o . \qquad (1.105)$$

The essential features of the dependence of W upon τ_o can be seen from the following simple argument: Provided $\tau_o << 1$ then $\tau << 1$ for all X and we may expand equation (1.102) so that

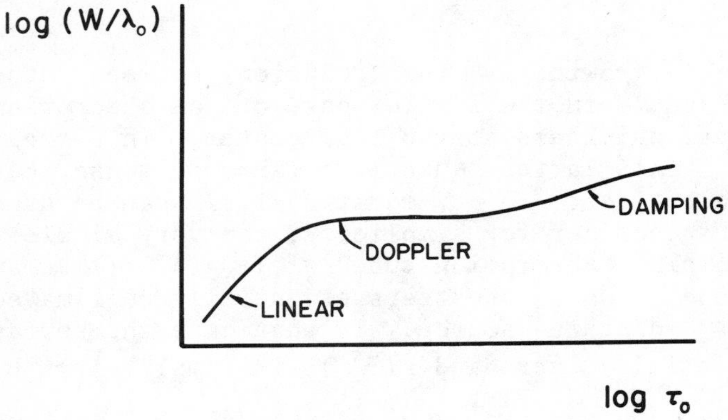

Fig. 1.6 The curve of growth

$$W \simeq \lambda_o\left(\frac{\ell}{v_D}\frac{\pi e^2}{mc}\right)(fm\lambda_o) \simeq \tau_o \quad (\tau_o << 1). \tag{1.106}$$

A weak line such as this is called unsaturated, and its equivalent width depends only upon its integrated strength and is independent of factors such as v_D or a which influence the shape of the absorption coefficient. As τ_o becomes greater r will diminish and for large τ is essentially zero so that $1 - r \simeq 1$. The value of r goes from <<1 to $\simeq 0$ so quickly that to obtain a crude idea of $W(\tau_o)$ we may replace (1.102) by

$$\begin{aligned} r(\Delta\lambda) &\simeq 0 \quad (\tau(\Delta\lambda)>1), \\ r(\Delta\lambda) &\simeq 1 \quad (\tau(\Delta\lambda)<1). \end{aligned} \tag{1.107}$$

Then we simply have

$$W \simeq \lambda_o \frac{v_D}{c} x^* , \tag{1.108}$$

where

$$\tau_o V(x^*,a) = 1. \tag{1.109}$$

For the core of the line, (see equation 1.89) $\tau \simeq \tau_o e^{-x^2}$ $(x \lesssim 3)$ or

$$x^* \simeq \sqrt{\ln\tau_o} . \tag{1.110}$$

For the damping wings, on the other hand (see equation 1.90)

$$x^* \propto \sqrt{\tau_0} \quad . \tag{1.111}$$

A sketch of log w/λ_o vs. log τ_0, in light of equations (1.106), (1.110) and (1.111) shows then that there are three main parts to the curve of growth: a linear portion, where the lines are weak and unsaturated, a "Doppler" or "flat" part where medium strength lines lie, and a damping part on which the stronger lines are formed (see Fig. 1.6).

The exact shape of the curve of growth clearly depends upon the relation between r and τ_0 (or r and η) and a comparison of equations (1.100) and (1.102) show that these differ: in principle every single line would have its own curve of growth, different for every stellar atmosphere, as one varied the abundance of the element responsible for the line. In practice, the general shape is dominated by the nature of V(x,a) and the general features, described by equations (1.106), (1.110), (1.111), remain the same though the details differ. Often, rather than calculate W theoretically, one chooses to let the star act as a giant analogue computer: one assembles many different lines from a given ion with a spread in f (but with the relative f-values known) which arise from a common set of levels so that n is the same (or can be compared with confidence) for each of the quantum states which form the lower level of the transition. Then, from the form of equation (1.104) one sees that a plot of log (W/λ_o) vs. log $(f\lambda_o)$ for each line will generate a curve of growth.

Comparison of the curve of growth of, say, lines of vanadium with those of say, iron, will give information on the V/Fe abundance ratio if: (a) the f-values of the iron lines relative to those of vanadium are known, and (b) differences in excitation and ionization are properly allowed for.

This idea can be extended to compare two stars of roughly similar type: one can also extract possible differences in v_D by this comparison. It should be noted that one really obtains, not simply abundance ratios between the two stars, since, as equation (1.100) makes clear, the continuous absorption coefficients enter into the line strengths, and if these differ from one star to another, this must also be properly allowed for.

2. Stellar Interiors

2.1. Interior Conditions

A star having spherical symmetry is built in a series of concentric spherical shells. Let M(r) be the total mass within a sphere of radius r; then the shell of thickness dr has the mass

$$dM(r) = 4\pi r^2 \rho dr, \tag{2.1}$$

where ρ is the mass density. The mass distribution in a spherical star depends only on how the density varies with distance from the center.

A stable star is in force balance, which means that the pressure must increase with depth fast enough to support the weight of the overlying layers. A mass element of unit cross section and height dr has weight

$$dw = g\ dm = \frac{GM(r)}{r^2} \rho dr\ . \tag{2.2}$$

where g is the acceleration of gravity at the given point. The pressure over dr must change by the same amount if force balance is maintained:

$$dP = -\frac{GM(r)}{r^2}\rho dr . \tag{2.3}$$

The minus sign indicates that pressure increases as distance decreases.

Equations (2.1) and (2.3) give the mass and pressure gradients within a stable star having spherical symmetry. These equations are not sufficient in themselves to determine the internal structure of a star, but they do put strong limits on that structure. The additional information needed to solve these equations comes from considering the production and flow of energy in stars.

Equation (2.3) can be applied in an order of magnitude fashion to determine the approximate conditions in a star. Let dP and dr be replaced by the finite increments corresponding to the step from the center to the surface of a star; then $dP = P_c - P_s$, $dr = r_c - r_s = -R$, where R is the radius of the star and the subscripts c and s refer to the center and the surface of the star. Granted that this is a very large step to take, if the other quantities in equation (2.3) are evaluated at some average value for the star, the result should be accurate to at least an order of magnitude. Neglecting the surface pressure as compared to its central value, one finds

$$P_c \simeq \frac{GM^2}{4R^4} = 3 \times 10^{15} \left(\frac{M^2}{R^4}\right)_o \text{ cgs} . \tag{2.4}$$

In the second part of equation (2.4) the mass and radius are given in solar units: $M_o = 2 \times 10^{33}$g, $R_o = 7 \times 10^{10}$ cm. The above not only indicates what general values of internal pressures to expect, but it also shows roughly how the pressure varies from one star to another of different mass and radius.

The mean density of the Sun is 1.4 g cm^{-3} . For any other star then,

$$\bar{\rho} = 1.4 \left(\frac{M}{R^3}\right)_o \text{ g cm}^{-3} . \tag{2.5}$$

The temperature can be introduced through the equation of state. In most cases the perfect gas law is quite accurate:

$$P = NkT = \frac{k\rho T}{m_o\mu} . \tag{2.6}$$

N is the number of free particles per unit volume, m_o is the mass of unit atomic weight (approximately the mass of the proton), and μ is the average atomic weight of the material. Substituting equations (2.5) and (2.6) into (2.4), one finds the internal temperature of a star to be roughly

$$T \simeq 10^7 \mu \left(\frac{M}{R}\right)_o \ {}^oK . \tag{2.7}$$

The atomic weight (also called the mean molecular weight) is of order unity, so temperatures around $10^{7\,o}$K should prevail within stars having roughly solar values of M and R. For stars whose interiors are highly degenerate, the perfect gas law is not valid and equation (2.7) may be greatly in error.

2.2. Radiation

In most stars the important modes of energy transport are radiation and convection. These mechanisms are discussed in the present and the next sections.

It is shown in Chapter 1 that the intensity of radiation I_ν satisfies the following equation of transfer:

$$\frac{dI_\nu}{ds} = k(S_\nu - I_\nu) , \tag{2.8}$$

where k is the volume absorption coefficient in cm^{-1}, S_ν is the source function, and ds is an element of distance along which the intensity is measured. If a star has spherical symmetry, it is convenient to express equation (2.8) in spherical coordinates (r,θ,ϕ):

$$\frac{1}{k}\left(\cos\theta\,\frac{\partial I_\nu}{\partial r} - \frac{\sin\theta}{r}\,\frac{\partial I_\nu}{\partial \theta}\right) = S_\nu - I_\nu \quad . \qquad (2.9)$$

The absorption coefficient k is at least as large as the contribution due to free electrons, which is numerically equal to several tenths times the density in cgs units. Thus the photon mean free path λ, which is $1/k$, is no greater than about 1 cm at most in typical stellar interiors. The terms in equation (2.9) must then approximately satisfy

$$\frac{\cos\theta}{k}\,\frac{\partial I_\nu}{\partial r} \simeq \frac{\lambda I_\nu}{R} << I_\nu \;, \qquad (2.10)$$

$$\frac{\sin\theta}{kr}\,\frac{\partial I_\nu}{\partial \theta} \simeq \frac{\lambda}{R}\,\frac{\partial I_\nu}{\partial \theta} << I_\nu \;. \qquad (2.11)$$

Each term of the left side of equation (2.9) is very small, which indicates that in stellar interiors the intensity is very nearly equal to the source function. Also the value of the mean free path means that a point is very effectively screened from any other point which is much more than 1 cm away from it. Physical conditions cannot normally vary much in 1 cm, so the radiation field in stellar interiors must be very close to local thermodynamic equilibrium: $S_\nu = B_\nu$, the Planck function. (cf. Chapter 1.)

One also has that the intensity is equal to the Planck function to a very good approximation, but the net flow of energy out of a star requires a finite anisotropy in the radiation field. Thus one can improve on this approximation in an important way by considering the two terms on the left side of equation (2.9) to be small perturbations which are to be evaluated under the approximation $I_\nu = B_\nu$.

The result is

$$I_\nu = B_\nu - \frac{\cos\theta}{k}\frac{dB_\nu}{dr} \quad . \tag{2.12}$$

The mean intensity, energy density, flux, and radiation pressure are important quantities introduced in Chapter 1 and derived from the intensity. One finds from equation (2.12):

$$\begin{aligned} J_\nu &= \frac{1}{4\pi}\int I_\nu \, d\omega = B_\nu \ , \\ u_\nu &= \frac{1}{c}\int I_\nu \, d\omega = \frac{4\pi}{c} B_\nu \ , \\ F_\nu &= \int I_\nu \cos\theta \, d\omega = -\frac{4\pi}{3k}\frac{dB_\nu}{dr} \ , \\ P_r(\nu) &= \frac{1}{c}\int I_\nu \cos^2\theta \, d\omega = \frac{4\pi}{3c} B_\nu \ . \end{aligned} \tag{2.13}$$

The flux depends directly on the small deviations from isotropy, while the other quantities are unaffected by the perturbation term in equation (2.12).

The integrated Planck function is given by

$$B = \int_0^\infty B_\nu \, d\nu = \frac{ac}{4\pi} T^4 \ , \tag{2.14}$$

where $a = (8\pi^5 k^4/15h^3c^3) = 7.56 \times 10^{-15}$ erg cm^{-3}°K^{-4} . From the above the integrated mean intensity, energy density, and radiation pressure can be related to the temperature:

$$J = B = \frac{ac}{4\pi} T^4 ,$$

$$u = \frac{4\pi}{c} B = aT^4 , \qquad (2.15)$$

$$P_r = \frac{4\pi}{3c} B = 1/3 \; aT^4 .$$

The flux is not so easily handled, as the absorption coefficient is a function of frequency.

It is common practice in stellar interior studies to introduce an average absorption coefficient or opacity k_o in terms of which the integrated quantities F and B have the same relation as do F_ν and B_ν in equation (2.13):

$$F = - \frac{4 \pi}{3k_o} \frac{dB}{dr} = - \frac{ac}{3k_o} \frac{dT^4}{dr} . \qquad (2.16)$$

This relation yields

$$- \frac{4\pi}{3k_o} \frac{dB}{dr} = \int_0^\infty F_\nu \, d\nu = - \frac{4\pi}{3} \int_0^\infty \frac{1}{k} \frac{dB_\nu}{dr} d\nu .$$

The common factor $(-4 \pi/3)$ can be cancelled from both sides. Also one can write $dB/dr = (dT/dr)(dB/dT)$, and the factor dT/dr can also be cancelled. The result is the following for the opacity:

$$\frac{1}{k_o}\frac{dB}{dT} = \int_0^\infty \frac{1}{k}\frac{dB_\nu}{dT}\, d\nu \; . \tag{2.17}$$

This form for the opacity is known as the Rosseland mean absorption coefficient. In terms of it equation (2.16) determines the radiative energy flow required by the gradient of T^4 .

2.3. Convection

Convection in addition to radiation can carry energy if conditions are right, and this will obviously modify the temperature gradient in a convective region of a star. First the conditions for the existence of convection will be established.

Consider within a star a mass element which is in complete equilibrium with its surroundings at a temperature T, pressure P, and density ρ. Now suppose the element is suddenly displaced upward by the amount dr, where its new surroundings have conditions T - dT, P - dP, and ρ - d ρ. For an instant the element will still have its original conditions, but its excess pressure will cause it to expand, reducing its pressure to P - dP, the same as the new surroundings. This expansion changes the temperature and density of the element to T - δT and ρ - $\delta\rho$. What happens now depends on the relative densities of the element and its environment: if the element is more dense than the region it finds itself in, it will sink back toward its original position and convective motions will be damped out. If the element is lighter than its surroundings, however, it will be buoyed up even further from the starting place. In this case conditions are such that disturbances in position are enhanced with time, and convection will occur.

The condition for convection is seen to be $(\rho - \delta\rho) < (\rho - d\rho)$, or $\delta\rho > d\rho$. If the element expands so rapidly that the energy exchange with the surroundings is negligible, it will take place adiabatically. One can then write the condition for convection as

$$\left(\frac{d\rho}{dr}\right)_{act} < \left(\frac{d\rho}{dr}\right)_{ad} \quad . \qquad (2.18)$$

Using the equation of state for a perfect gas, one can express this in terms of the temperature:

$$\frac{d}{dr}(T/\mu)_{act} > \frac{d}{dr}(T/\mu)_{ad} \quad . \qquad (2.19)$$

Convection will occur if the density gradient is less than the adiabat, or equivalently, if the temperature gradient is greater than the adiabat. The same relations would have followed if the element had been displaced downward.

If a region in a star does have convective motions, the energy flux carried by these motions is

$$F_c = c_p \rho v \delta T \ , \qquad (2.20)$$

where c_p is the specific heat per gram at constant pressure, v is the average velocity of a convective element, and δT is the typical temperature excess or deficit of a convective element compared to its surroundings. The convective velocity can be estimated as follows:

The buoyant force on an element is $g\,\delta\rho$, where g is the local acceleration of gravity. If an element rises or sinks through a distance s before it dissolves into the surroundings, then the work done per unit volume on an element chosen at random is approximately

$$W = \int_o^{s/2} g\delta\rho \; dr = \int_o^{s/2} \frac{g\rho\,\delta T}{T} \, dr \quad . \qquad (2.21)$$

The integral is carried out over the path of the convective element. If the element moves adiabatically,

$$\delta T = \left| \left(\frac{dT}{dr}\right)_{ad} - \left(\frac{dT}{dr}\right)_{act} \right| \delta r. \tag{2.22}$$

Using this in equation (2.21) and equating W with $1/2\rho v^2$, the convective velocity is

$$v^2 = \frac{gs^2}{4T} \left| \left(\frac{dT}{dr}\right)_{ad} - \left(\frac{dT}{dr}\right)_{act} \right| . \tag{2.23}$$

The convective flux of equation (2.20) is then

$$\mathcal{F}_c = c_p \rho \left(\frac{g}{T}\right)^{1/2} \frac{s^2}{4} \left| \left(\frac{dT}{dr}\right)_{ad} - \left(\frac{dT}{dr}\right)_{act} \right|^{3/2} . \tag{2.24}$$

The inequality (2.19) implies that the convective flux vanishes if the actual temperature gradient is less than or equal to the adiabatic gradient. The distance s is known as the mixing length of the convective motions.

The analysis that led to equation (2.24) is hardly more accurate than an order of magnitude, so the numerical factor should not be considered well determined. As will be seen, this is not generally a serious shortcoming in stellar interiors.

The mixing length s can be expected on rather general arguments to be of the order of size of the scale height, i.e., the distance over which the physical variables vary significantly. Using equation (2.3) for the pressure gradient, it is seen that the distance over which the pressure varies by an amount comparable to itself is roughly

$$s \simeq \frac{r^2 P}{GM(r)\rho} \simeq 10^{10} \ (R)_o \ \text{cm} , \tag{2.25}$$

where as usual $(R)_o$ means the radius is to be measured in solar units. Putting in other numerical values, including equation (2.7), equation (2.24) is

$$F_c \simeq 2 \times 10^{26} \left| \left(\frac{dT}{dr}\right)_{ad} - \left(\frac{dT}{dr}\right)_{act} \right|^{3/2} \left(\frac{M}{R^{3/2}}\right)_o . \quad (2.26)$$

The total flux of energy F(r) due to both radiation and convection is given by $L(r)/4\pi r^2$, where L(r) is the luminosity at distance r from the center. One then obtains for the relative excess temperature gradient in a convective region within a star

$$\left| \frac{\left(\frac{dT}{dr}\right)_{ad} - \left(\frac{dT}{dr}\right)_{act}}{\left(\frac{dT}{dr}\right)_{act}} \right| \simeq 5 \times 10^{-7} \left(\frac{F_c}{F}\right)^{2/3} \left(\frac{R^5 L^2}{M^5}\right)_o^{1/3} . (2.27)$$

The convective flux can be no larger than the total flux, so the above shows that the temperature gradient in a convective zone exceeds the adiabatic by a negligible amount. One is justified in assuming that the physical variables follow the adiabat when convection is in force, and that convection will carry however much energy it needs to in order to account for the luminosity of the star.

In the atmosphere of a star the situation is quite different. The density of matter and radiation is so low that convective motions do not transport energy efficiently. The result is that the temperature gradient can be much in excess of the adiabatic without convection carrying much of the total energy.

2.4. Gas Laws

A perfect gas is one that obeys the Maxwell-

Boltzmann statistics. The velocity distribution of such a gas is

$$p(\vec{v})\ d\vec{v} = \left(\frac{m}{2\pi kT}\right)^{3/2} \exp-\left(\frac{mv^2}{2kT}\right) d\vec{v} \quad . \tag{2.28}$$

Here v is the magnitude of the three dimensional vector $\vec{v}$, and equation (2.28) gives the probability that a gas particle has its velocity in the given range. The average kinetic energy of a particle is

$$\overline{KE} = \int 1/2mv^2 p(v)\ dv = \frac{3}{2}kT \quad . \tag{2.29}$$

The pressure of a perfect gas can also be found from the above. In time interval dt a particle will move the distance $v_x dt$ in the x direction. If N is the number of gas particles per unit volume, then $Ndydzv_x dtp(v_x)dv_x$ is the number of particles having velocity components between v_x and $v_x + dv_x$ and which cross the area dydz in the interval dt. $p(v_x)dv_x$ is the expression (2.28) integrated over all values of the y and z components of velocity. Multiplying the above quantity by mv_x gives the momentum carried across the surface. Integrating the result over all values of v_x and evaluating it per unit area and time gives the pressure:

$$P = \int_{-\infty}^{+\infty} mv_x Nv_x p(v_x) dv_x = NkT \quad . \tag{2.30}$$

This was used in equation (2.6). The alternate form follows by expressing N as $(\rho /m) = (\rho/m_o \mu)$. m is the mass per particle and $m_o = 1.660 \times 10^{-24}$ g is the mass of unit atomic weight. If the gas is a mixture of different kinds of particles, equation (2.30) is still valid if N is the total number of particles per unit volume. Then m is the average mass per free particle and μ is the average molecular weight of the mixture. Changing ionization can obviously make both m and μ functions of the physical conditions.

The main cause of deviations from the perfect gas law in stellar interiors is degeneracy. Degeneracy is due to the exclusion principle, which is the statement that no two identical particles of half integral spin can exist in the same unit of volume and in the same quantum mechanical state. At relatively low densities the number of available states is far larger than the number of particles which can occupy them, so the exclusion principle has no practical effect; the perfect gas laws are valid. At sufficiently high densities, however, a significant fraction of the quantum states will be filled and deviations from the perfect gas laws will result.

The quantum nature of a particle is apparent on a spacial resolution of the order of the deBroglie wavelength λ_D. Thus the density of particles necessary to make the exclusion principle important is one in which the mean distance between particles is of the order of the deBroglie wavelength. To put it the other way around, if the mean distance between particles is much larger than the deBroglie wavelength, degeneracy should be unimportant:

$$\bar{d} = \left(\frac{3}{4\pi N} \right)^{1/3} >> \lambda_D = \frac{h}{(2mKE)^{1/2}} \quad . \tag{2.31}$$

If this inequality is satisfied, the average particle kinetic energy is given by equation (2.29). Using this for KE, inequality (2.31) becomes for no degeneracy

$$T >> \frac{h^2}{mk} N^{2/3} = 3 \times 10^{-37} \frac{N^{2/3}}{m} \quad . \tag{2.32}$$

This relation must be applied to each type of particle separately. Electrons with their small mass become degenerate at much lower particle densities than do the heavier ions.

In a degenerate gas the velocity distribution is given by Fermi-Dirac relation

$$p(\vec{v})\ d\vec{v} = \frac{2m^3}{Nh^3}\ \frac{d\vec{v}}{\exp(\alpha+mv^2/2kT) + 1}\ ; \qquad (2.33)$$

α is a normalizing quantity that depends on N and T. If $\alpha \gg 1$, then the unity in the denominator is negligible and equation (2.33) reduces to the non-degenerate expression (2.28). The smaller the value of α, the larger is N at fixed temperature, and the more important is degeneracy. Note that degeneracy depends upon both density and temperature. The higher the temperature, the larger the density must be for degeneracy to set it.

The Fermi-Dirac relation (2.33) naturally leads to a different equation of state than the classical expression. A degenerate gas has a higher pressure and a higher energy than predicted by the classical relations. The reason is that the particles have some difficulty in finding empty quantum states and, as a consequence, are forced to higher energy levels than would otherwise be the case. Another important point is that as degeneracy becomes more nearly complete, the energies of the particles become determined simply by what energy levels are available, and the temperature becomes a less relevant quantity.

Consider the case of complete degeneracy: all quantum states are completely filled up to a certain level. This corresponds to $\alpha \to -\infty$ in equation (2.33), or

$$N\ p(\vec{v})\ d\vec{v} = \frac{2m^3}{h^3}\ d\vec{v} \qquad \vec{v} \leq \vec{v}_o\ , \qquad (2.34)$$

$$N = \frac{8\pi m^3}{3h^3}\ v_o^3\ , \qquad (2.35)$$

where v_o is the speed corresponding to the highest filled energy level. Following equation (2.30) for the pressure determination, but using the Fermi-Dirac distribution, one finds

$$P = \frac{8\pi m^4}{15h^3} v_o^5 . \qquad (2.36)$$

Eliminating v_o between the above two equations, one has the equation of state

$$P = \frac{1}{20}\left(\frac{3}{\pi}\right)^{2/3} \frac{h^2}{m} N^{5/3} . \qquad (2.37)$$

The mean kinetic energy is

$$\overline{KE} = \frac{3}{2}\frac{P}{N} . \qquad (2.38)$$

As stated above, in complete degeneracy the pressure and the energy are independent of the temperature.

It often happens that degeneracy forces the particles into such high energy states that their velocities are relativistic. In that case the above relations need a further modification, but the principle of degeneracy is the same: as the density increases at a fixed temperature, the pressure and energy of the particles become less dependent upon the temperature and more dependent on the placement of the allowed quantum states. When the quantum states are completely filled up to a certain level, no additional particles of the same type can be added unless their energies are above the threshold level.

2.5. The Virial Theorem

Multiply the hydrostatic equilibrium relation (2.3) by r/dr and integrate the result over the volume of a star:

$$\int_0^R \left[r\frac{dP}{dr} + \frac{GM(r)\rho}{r} \right] 4\pi r^2 dr = 0 . \qquad (2.39)$$

In the second term above replace the integration variable dr with dM through equation (2.1), and integrate the first term above by parts. If the surface pressure is negligibly small, the result is

$$\int_0^R 12\pi P\, r^2 dr = \int_0^M \frac{GM(r)\,dM}{r} \quad . \tag{2.40}$$

The term on the right is $-E_g$, the negative of the gravitational energy of the star. The pressure can be expressed as the sum of the gas and the radiation pressures. Assuming the perfect gas law for P_g and equation (2.15) for P_r, equation (2.40) yields

$$-E_g = 12\pi \int_0^R \left(\frac{kT\rho}{m} + \frac{1}{3}aT^4 \right) r^2 dr = \int_0^R (2u_{th} + u_r)\, 4\pi r^2 dr,$$

where u_{th} and u_r are the thermal and radiation energies per unit volume. By integrating them over the volume of the star, one obtains

$$2E_{th} + E_r + E_g = 0 \;, \tag{2.41}$$

where the E's are the given types of energies for the whole star. Equation (2.41) is known as the virial theorem; it is a constraint on how the energy in a stable star can be distributed among the various types.

The values of E_{th}, E_r, and E_g depend to some extent on the detailed internal structure of a star; however, their approximate values can be found by calculating them from a simplified model having constant density. For such a model,

$$E_{th} = 0.3G\,\frac{M^2}{R} = 1.1 \times 10^{48}\left(\frac{M^2}{R}\right)_o \text{ erg} \;, \tag{2.42}$$

$$E_r = \frac{32\pi a}{3465}\left(\frac{m_o \mu G}{k}\right)^4 \frac{M^4}{R} = 2.0 \times 10^{46} \mu^4 \left(\frac{M^4}{R}\right)_o \text{erg}, \tag{2.43}$$

$$E_g = -\frac{3}{5}\frac{GM^2}{R} = -2.3 \times 10^{48} \left(\frac{M^2}{R}\right)_o \text{ erg.} \tag{2.44}$$

Note that for a solar type star the radiation energy is only a few percent of the thermal energy. E_r varies as the fourth power of the mass, however, while E_{th} is proportional to the square of the mass; thus for high mass stars the radiation energy become more significant.

The thermal energy of a star is the direct source of its luminosity. If a star cannot replenish its thermal energy in some way, its lifetime will be limited to about

$$t_{th} \simeq \frac{E_{th}}{L} \simeq 2.9 \times 10^{14} \left(\frac{M^2}{LR}\right)_o \text{ sec.} \tag{2.45}$$

This is about 10^7 years for the Sun, far younger than the known age of the Sun (and most other stars). It follows that the thermal energy of the Sun is being replenished from some other source. One might suspect the gravitational energy as this source, but it is apparent from equations (2.42) and (2.44) that E_g is only of the same order of size as E_{th}. A much larger source is needed, and in the next section it will be shown that the nuclear sources are sufficient.

Define E as the sum of the thermal, radiation, and gravitational energies of a star:

$$E = E_{th} + E_r + E_g \ . \tag{2.46}$$

The virial theorem is

$$E + E_{th} = 0. \tag{2.47}$$

A non-zero value of E_{th} means that the star is losing energy through its luminosity. If no other form of energy, such a nuclear energy, becomes involved, the luminosity must result in a decrease in E:

$$L = -\frac{dE}{dt} = \frac{dE_{th}}{dt} . \qquad (2.48)$$

This is an interesting result: the lost energy actually causes the thermal energy to increase, i.e., the star gets hotter. It is noted through equation (2.42) that the increase in thermal energy must come from a decrease in radius, as the mass is assumed to be constant. The decrease of radius releases gravitational energy according to equation (2.44), and this provides for both the luminosity and the increased temperature.

From the above we can find the rate of contraction needed to support a given luminosity. It is

$$\frac{dR}{dt} = -\frac{10R^2L}{3GM^2} = -2.4 \times 10^{-4}\left(\frac{R^2L}{M^2}\right)_o \text{ cm sec}^{-1} . \qquad (2.49)$$

A shrinking of the solar radius by this amount could not be detected, but the lifetime argument mentioned above leaves no doubt that the Sun is not deriving its luminosity from contraction.

An original contraction is the method by which stars get hot in the first place. Contraction is also important between different nuclear burning stages in the lives of stars. Note that, classically, there is no limit to how much gravitational energy a star can release. Equation (2.44) indicates how much energy has been released in the contraction from infinite size down to radius R. As R goes to zero, E_g goes to minus infinity, although degeneracy may not allow unlimited contraction.

2.6. Nuclear Energy

The lifetimes of the stars depend directly upon the amount of nuclear energy which they have available for conversion into thermal energy. Nuclear fusion, the building up of heavy nuclei from lighter ones, is the main mechanism of nuclear energy release in stars.

Suppose nuclear reactions occur which change element i into element j. If n is the ratio of mass numbers of the relevant isotopes of the elements, then one complete reaction will convert n nuclei of type i into one nucleus of type j. The energy release by this is $(nm_i - m_j)c^2$. If this is divided by nm_i, one obtains the available energy from this type of reaction per gram of element i. Finally, by multiplying this result times $x_i M$, where x_i is the abundance fraction of element i by mass, we obtain the energy available to a star by this reaction:

$$E_n = \frac{(nm_i - m_j)c^2}{nm_i} x_i M. \tag{2.50}$$

Hydrogen is both the lightest and the most abundant element in most stars, so it is a reasonable choice for element i. Iron nuclei have the strongest binding, which means that the most efficient energy release is obtained by converting other elements into iron. One can find an upper limit to the available nuclear energy in a star by assuming that the star starts as pure hydrogen and ends up pure iron. For this case $m_i = 1.0078m_o$, $m_j = 55.9353m_o$, $n = 56$, and $x_i = 1$; the result is

$$E_n(H \rightarrow Fe) = 1.59 \times 10^{52} (M)_o \text{ erg.} \tag{2.51}$$

Comparing this value with equation (2.42), we see that the nuclear energy supply is nearly four orders of magnitude greater than the thermal energy. As long as equation (2.51) is not too generous an upper limit, nuclear energy can account for the known ages of the Sun (about 5×10^9 years) and of other stars (up to about 10^{10} years).

The above may be questioned from several directions. First, the nuclear reactions may be halted before iron is reached in the chain of fusion processes. Second, not all the material in a star may be available.

For the first point, the conversion of hydrogen into iron requires many steps, the first of which is hydrogen into helium. As will be seen later, some stars never get past this first step. It is important to find out how this affects stellar lifetimes. For helium production

$m_j = 4.0026 m_o$, $n = 4$:

$$E_n(H\to He) = 1.27 \times 10^{52} (M)_o \text{ erg} = 0.8\, E_n(H\to Fe). \quad (2.52)$$

80% of the available nuclear energy release occurs in the first step of helium production. The later steps do not significantly extend the lifetimes of the stars.

For the second point, the reactions are strongly concentrated toward the centers of the stars, where the temperatures and densities are greatest. The outer regions may never become involved, thus cutting down on the amount of available fuel. The following argument limits the importance of this effect: if the nuclear reactions are spread over a moderately large region, a moderately large fraction of the mass of the star will be involved; if they are concentrated in a very small region at the center, a very large temperature gradient will be needed to carry the energy away, and this will cause the central regions to be convective (equation 2.19). Convection of course will bring in fuel from a much larger region, again making a significant fraction of the whole mass of the star involved. The result is that no less than about 10% of the mass of a star becomes involved in the nuclear reactions. Equation (2.51) may be up to about one order of magnitude larger than a realistic value of the available nuclear energy.

The most important reactions are those which convert hydrogen to helium. There are two main sets of reactions along which this can be carried out:

$$\begin{aligned} &2(H^1 + H^1) \to 2(H^2 + \beta^+ + \nu), \\ &2(H^2 + H^1) \to 2(He^3 + \gamma), \\ &He^3 + He^3 \to He^4 + 2H^1; \end{aligned} \quad (2.53)$$

$$\begin{aligned} &C^{12} + H^1 \to N^{13} + \gamma, \\ &\qquad N^{13} \to C^{13} + \beta^+ + \nu, \\ &C^{13} + H^1 \to N^{14} + \gamma, \end{aligned} \quad (2.54)$$

$$N^{14} + H^{1} \rightarrow O^{15} + \gamma,$$

$$O^{15} \rightarrow N^{15} + \beta^{+} + \nu,$$

$$N^{15} + H^{1} \rightarrow C^{12} + He^{4}.$$

In the above, β^{+} represents a positron, ν is a neutrino, and γ is a gamma ray photon. Reactions (2.53) make up what is called the proton-proton (PP) reaction, while (2.54) is the carbon-nitrogen cycle (CN). The former produces helium directly from hydrogen, while the latter uses carbon and nitrogen as catalysts. A pure hydrogen star of course could not use the CN cycle, but in a star with moderate amounts of the heavy elements both sets of reactions (2.53) and (2.54) would compete with each other.

The neutrinos produced in these reactions escape from the stars without interacting with other particles. The energy they escape with is not generally included in the luminosity of a star, so the neutrino energy must be subtracted from the available nuclear energy before comparisons with luminosity are meaningful. This is only a small correction in the hydrogen burning processes, but neutrinos can provide an important mechanism of energy loss in later stages in the lives of some stars.

The main barrier to nuclear reactions is the Coulomb repulsion between two nuclei. This makes it very difficult for two nuclei to get close enough together to cause a reaction. The PP chain involves only light particles with small nuclear charges, so the Coulomb barrier is relatively small; the CN cycle uses heavier nuclei with much larger Coulomb barriers. At lower temperatures the particle energies are not sufficient to significantly penetrate the stronger barriers, and the PP reaction will dominate. The CN reactions are much more sensitive to the temperature, and at sufficiently high temperatures they dominate.

The reaction $(H^1 + H^1)$ is by far the most difficult to bring about of the PP reactions, and $(N^{14} + H^1)$ is the slowest of the CN cycle. For the CN cycle, practically all of the carbon and nitrogen near the center of a star will be converted to N^{14} while waiting for the N^{14} to react with a proton.

It is often convenient to express the rate of nuclear energy release in terms of powers of density and temperature. For the reactions considered above, the energy release in ergs per gram per second are approximately:

$$\varepsilon_{PP} = \varepsilon_o(PP)\, x_H^{\,2} \rho T^4 \quad , \tag{2.55}$$

$$\varepsilon_{CN} = \varepsilon_o(CN)\, x_H x_{CN} \rho T^{16} \quad . \tag{2.56}$$

These relations are approximately valid for the temperature regions of importance. There are many other nuclear reactions which can be important in stars, and each will have its own dependence on physical conditions.

2.7. The Russell-Vogt Theorem

The equations governing the structure of a stable star are:

$$\frac{dM(r)}{dr} = 4\pi r^2 \rho \quad , \tag{2.57}$$

$$\frac{dP}{dr} = - \frac{GM(r)\rho}{r^2} \quad , \tag{2.58}$$

$$\frac{dL(r)}{dr} = 4\pi r^2 \rho \varepsilon \quad ; \tag{2.59}$$

ε is the thermal energy source per gram-second. It will include the conversion of gravitational energy if the star is contracting. The luminosity carried by radiation is

$$L_r(r) = -\frac{4\pi acr^2}{3\,k_o}\,\frac{dT^4}{dr}\ . \tag{2.60}$$

In a convection zone, the adiabatic gradient prevails unless the zone is very close to the surface:

$$\frac{dT}{dP} = \left(\frac{dT}{dP}\right)_{ad}\ . \tag{2.61}$$

We now investigate the solution of the above equations to find the internal structure of a model star. It is assumed that we know enough physics to be able to calculate the energy generation, the opacity, and the adiabatic gradient for any given set of physical conditions. It is apparent that these quantities will depend upon the chemical composition of the material, so this composition must be given ahead of time.

Suppose T_c and P_c, the central values of the temperature and the pressure are assumed. Then the central value of the density follows from the equation of state, and the right sides of equations (2.57)-(2.59) are known at $r = 0$. This allows $M(r)$, P, and $L(r)$ to be found at a point slightly away from the center. The temperature at this new point follows from T_c and equation (2.60) if there is no convection, or with equation (2.61) if there is. Thus T and P are found at the new point, and this process is continued in small increments of r throughout the model star.

The surface of the star is reached when the mass is used up and the density goes to zero; however, this condition is not sufficient. The outer layers must radiate energy into space at the same rate they receive it from below, which means that a self-consistent model atmosphere must be appended to the model interior. (Stellar atmospheres are the subject of Chapter 1.) For present purposes, this means that the temperature must approach a calculable surface value T_o at the same point that the density goes to zero:

$$T \to T_o \text{ as } \rho \to 0 \,. \tag{2.62}$$

For some purposes setting $T_o = 0$ is sufficiently accurate, or it can be more accurately equated to the effective temperature (see Chapter 1). The main point is that there are two conditions (2.62) which must be satisfied at the surface, and in general they cannot both be satisfied at a single value of r. This means that a physically realistic model of a star cannot be constructed for arbitrary values of the central temperature and pressure. One can go back to the beginning, adjust P_c, and start the integration over again. In this fashion a value of P_c would eventually be found for which both conditions (2.62) would be satisfied at the same radius: a model star is obtained.

The complete structure of the model star has been determined, including the mass, radius, and luminosity. The only input parameters which were assumed at the start are the central temperature and the chemical composition. It is apparent, however, that we could have started with the value of any other parameter besides T_c, and the result would have been the same. For example, one could have assumed R to be known from the start. Then the conditions (2.62) would have to be satisfied specifically at $r = R$. Both T_c and P_c would need to be adjusted to satisfy this, and again a unique structure would be the result.

From the above it is seen that the complete structure of a star is determined by its composition plus one other parameter. That other parameter is usually chosen as the mass, since the mass is in many respects the most significant property of a star. The statement that mass and composition fix the complete structure of a star is known as the Russell-Vogt theorem.

As the star ages, nuclear reactions change the composition in its interior, while the outside layers remain unaffected. This does not matter as far as the Russell-Vogt theorem is concerned, for it was assumed that the composition be known at each point, not that it be uniform. The internal composition is not available to observation, but it can be calculated for a star of known age. Thus it is perhaps more useful to rephrase the Russell-Vogt theorem: the structure of a star is determined by its

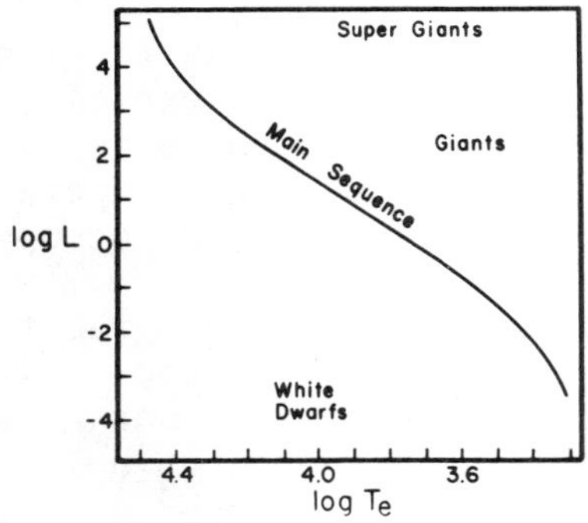

Fig. 2.1 The H-R diagram

mass, age, and original composition.

The Russell-Vogt theorem is not rigorously true for a number of reasons. First, there are other effects which can influence a star and which have been ignored here. Examples are angular momentum, magnetic field, and interaction with nearby stars or interstellar matter. Second, there are phases in the lives of stars in which they become unstable. While the equations considered here should be sufficient to predict the occurrence of these instabilities, they cannot predict the results of them. Third, a number of cases have been discovered in which more than one stable model corresponds to the same mass and distribution of composition. In these cases only one model lies on the evolutionary path of a stable star, so the other configurations can be reached only by stars which pass through significant instabilities. The Russell-Vogt theorem does have its qualifications, but it is an extremely important constraint on stable stellar configurations.

2.8. Stellar Evolution

Figure 2.1 is a Hertzsprung-Russell or HR diagram. This is a plot of some function of luminosity as ordinate and another function of surface temperature as abscissa, with the temperature traditionally increasing to the left. A star is represented by a point in the diagram. It is convenient and useful to classify a star by its position in the HR diagram.

Most stars are observed to be on the main sequence, a line running diagonally from the upper left to the lower right side of the HR diagram. Cool, faint stars on the lower end of the main sequence are much more common than those which are hotter and brighter. Giants are rather rare stars of fairly high luminosity which occupy a nearly horizontal strip in the HR diagram that extends to the right of the upper main sequence. Supergiants are extremely rare stars of extremely high luminosity, while white dwarfs are very low luminosity stars of moderately high temperature. There are many other types of stars which occupy special regions of the HR diagram.

How are the observed features of the HR diagram to be interpreted? Model calculations indicate that the main sequence is composed of stars which are homogeneous in composition and which are converting hydrogen into helium at a rate that exactly balances the luminosity. When a star first gets hot enough inside to ignite hydrogen in a significant way, it will settle into energy balance as a main sequence star. The position on the main sequence is determined by the mass: high mass means a hot, luminous star, and low mass leads to a cool, faint star. Differences in original composition among stars are relatively small and do not produce major effects, although they are quite important in many detailed studies.

Main sequence stars are too young to have appreciably changed their compositions through nuclear reactions. Eventually they will use up nearly all of their central hydrogen and they will be at least temporarily out of nuclear fuel. How long will this take? Equation (2.51) shows that the Sun can produce over 10^{52} ergs by converting its hydrogen to helium. Only the inner 10% or so of the mass is immediately available, however, so this number drops to 10^{51} erg. The solar luminosity is 4×10^{33} erg/sec, so the central hydrogen supply should last about 3×10^{17} sec $= 10^{10}$ yr at the present rate of consumption. The Sun can be expected to last about 10 billion years as a main sequence star. For a star of different mass and luminosity, the main sequence lifetime is

$$t_{ms} = 10^{10}\left(\frac{M}{L}\right)_{o} \text{ yr.} \qquad (2.63)$$

The most luminous stars on the main sequence have perhaps 10^5 times the luminosity of the Sun, but their mass is only about 50 M_o. They will last on the main sequence only a few million years. At the other end of the main sequence are stars having 10^{-4} L_o and about $0.1 M_o$. Such stars can shine for 10^{13} years before their central hydrogen supply is depleted. The smaller the mass of the star, the more conservative it is in using up its energy supply. Small mass stars have much longer lifetimes than the high mass stars. The range in mass among stars is much smaller than the ranges of most other stellar properties.

What happens when the central hydrogen is exhausted? All particles can undergo nuclear reactions, but the heavier nuclei have larger Coulomb barriers to overcome and require higher temperatures before they can react. When the core hydrogen is used up, the star is temporarily out of fuel and must contract. The consequent release of gravitational energy will both provide for the luminosity and make the star hotter. Eventually regions outside the exhausted core will become hot enough for hydrogen burning. Reactions will then take place in a shell around the helium core.

When the hydrogen in the shell is used up, further contraction will take place, heating the interior even more and igniting the hydrogen even further out from the center. In this way the hydrogen burning shell will eat its way through the star. Eventually contraction will heat up the core enough for the helium to be ignited, and helium burning reactions which produce carbon will supplement the hydrogen burning shell.

During the successive stages of contraction and shell burning, the luminosity increases and the outer layers expand. While the inner regions contract and grow hotter, the outer layers of the star actually cool off and the star moves upward and to the right in the HR diagram - into the region of the cool giants.

In the general manner described above, a star can alternate contraction and shell burning of different kinds of nuclear fuels until the core is hot enough for igniting the next set of nuclear reactions. It will not take long for a new fuel to be exhausted because, as noted earlier, most of the available energy is released during hydrogen burning, and because these later phases usually occur at high luminosity. The star will become a series of concentric shells of different composition, the smaller shells being made of heavier material. This process will continue until stopped by degeneracy or by instabilities.

As stated in Section 2.4, degeneracy occurs when the density is above a certain limit, and that limit increases as the material gets hotter. When a star contracts, the density goes up but so does the temperature. Does the density increase fast enough to catch up to the critical value needed for degeneracy? The answer depends on the mass of the star. Below about 1.4 M_o, known as the Chandrasekhar limit, the electrons in a star can become completely degenerate. Above that limiting mass, contraction increases the temperature fast enough to keep the critical density beyond reach: the electrons cannot become completely degenerate no matter how much the contraction.

The white dwarfs are stars whose extremely high densities indicate electron degeneracy. The heavier ions are not degenerate, but don't contribute significantly to the pressure. All the known white dwarfs have masses well below the Chandrasekhar limit.

Instabilities which can alter the direction of evolution of a star occur in different forms. These can be relatively mild or quite violent, and they often involve the loss of mass by the star. For example, many white dwarfs have less than a solar mass, yet a star which had this small a mass when formed probably could not be old enough to have evolved to the white dwarf stage. (Star formation in the Galaxy began only some 10^{10} years ago.) These low mass white dwarfs must have been formed with considerably more mass, and they lost some of it through later instabilities. Significant mass loss by stars must be quite common, and many types of stars are observed to be losing mass. Planetary nebulae are thought to be the immediate predecessors of white dwarfs.

Rotational instability in a contracting star can cause mass loss, as can radiation pressure, particularly in a cool giant or supergiant with a highly extended outer envelope. More explosive situations also occur. Supernovae are stars which are observed to flare up in brightness by a factor of 10^9 or so in a matter of hours, and they fade away over the following several months. Supernovae are obviously caused by extreme instabilities.

If the core of a contracting star is highly degenerate and if a new element in the core is suddenly ignited in nuclear reactions, the situation can be very explosive. The reactions suddenly release large amounts of energy, but the degenerate electrons, which supply practically all of the pressure, are not affected. The released energy heats up the nuclei without the safety valve action of expansion of the core, and the result is a runaway in the nculear reaction rate. Suddenly degeneracy is removed, the pressure skyrockets, and the core is far out of pressure-gravity balance. This flash, as it is called, that occurs when a new element is ignited in a degenerate core may or may not be violent enough to trigger a supernova explosion.

Another possible cause of supernovae is an iron core. If the alternating processes of contraction and nuclear reactions have built up a core of iron, any further reactions absorb energy from the surroundings rather than release energy to it. Contraction of an iron core at extreme temperatures and densities causes the iron to break up into helium and smaller particles with the absorption of large amounts of energy. The sudden drain of thermal energy throws the core far out of pressure-gravity balance, and a major collapse occurs. The collapse would compress and ignite the lighter material surrounding the core, and a supernova explosion could possibly result.

The Crab nebula is the gas ejected in a supernova explosion over 900 years ago. The star which had this explosion has been identified as a pulsar which emits nonthermal radiation in pulses every 0.03 seconds. Pulsars are almost certainly neutron stars, stars that are so dense that most of the surviving particles are neutrons. Neutron degeneracy is a possible final state for stars, being analogous to the electron degeneracy of white dwarfs, but much more extreme. Other pulsars have been identified with old supernovae, so there is considerable direct evidence that the remnant of a supernova outburst is a neutron star.

As in the case of white dwarfs, there is an upper limit to the mass of a neutron star which can become completely degenerate and halt the contraction. What about a star that is too massive to become degenerate at all? Must it necessarily lose enough mass to come below the degeneracy limit? Present theory accepts the possibility that such a star can contract to the degree that its surface gravity allows neither matter nor radiation to escape from it: a black hole is formed. Whether black holes exist is not known at present, but one component of the binary x ray source called Cygnus X-1 is considered to be a plausible candidate.

Star clusters provide an important source of observations for stellar evolution studies. The reason is that the members of a given cluster are probably of about the same age and made out of the same kind of material. The HR diagram of a given cluster shows the effects of different masses on the properties of stars, and age effects can be found by comparing different clusters. There is the

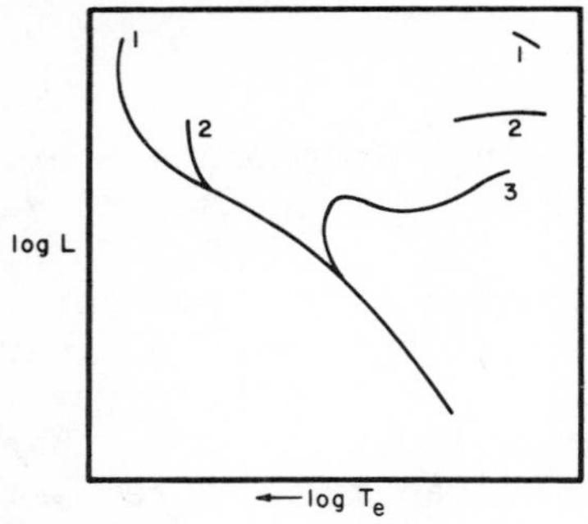

Fig. 2.2 Cluster H-R diagrams

additional point that all members of a cluster are at essentially the same distance from us. Important information can be obtained from the relative brightnesses of the stars even if the distance to the cluster is not known, i.e., even if the luminosities of the stars cannot be found accurately.

Figure 2.2 shows schematic HR diagrams of three hypothetical clusters. Cluster 1 is almost pure main sequence plus a small number of cool supergiants. This is what is expected of a very young cluster, as only the most massive stars have had time to evolve off of the main sequence. Cluster 2 has the same main sequence as cluster 1 below a certain turn off point, but above that point it curves a little above and to the right of the cluster 1 main sequence. Also cluster 2 has some cool giants, more in number and lower in luminosity than the cool supergiants of cluster 1. Cluster 2 is older than 1, and stars of lower mass have had time to exhaust their core hydrogen and evolve away from the main sequence. Stars of very large mass, which are the supergiants of cluster 1, have long since burned themselves out in cluster 2 and by now are white dwarfs, neutron stars, or black holes. Cluster 3 has a still lower turn off point and its giants have still lower luminosities. Only its relatively low mass stars are still on the main sequence. This is the oldest of the three clusters. Composition differences exist between clusters, particularly between the oldest globular clusters and the younger open clusters, and these differences must be taken into account in detailed comparisons.

The earliest and latest stages in the lives of stars - the formation of stars from interstellar clouds and the mechanisms of instability and mass loss in post main sequence phases - are the least understood parts of stellar evolution. The discrepancy between the observed and calculated rates of neutrino production by the Sun, however, shows that even main sequence stars can present important unsolved problems.

3.

Gaseous Nebulae

3.1 Introduction

Large tenuous clouds of ionized gas which are associated with hot stars, and which emit visible light because of the energy they receive from the ultraviolet radiation of the stars are called gaseous nebulae. Such emission nebulae are to be distinguished from bright nebulae which shine because of the reflection of starlight by dust particles (reflection nebulae), and from dark nebulae, which are caused by the obscuration of stellar radiation by intervening interstellar grains.

The primary constituents of the interstellar medium may be divided into two types of material: gas, and microscopic solid particles called grains (or dust). The gas emits radiation in the optical region of the spectrum by absorbing the far-UV radiation from stars and converting it to visual wavelengths. The visual luminosity of grains, on the other hand, derives from their simple scattering of the visual radiation of nearby stars. As a result, the relative brightness of a gaseous emission nebula as compared with a reflection nebula around a star is determined by two factors: (a) the dust/gas ratio, and (b) the ratio of the far-UV to visual luminosity of the star. Because of the great sensitivity of the UV flux emitted by a star to the temperature of its surface layers, the most important criterion for the existence of a gaseous nebula around a star is that the star be hot. Hubble discovered the empirical law that gaseous nebulae occur only around stars of spectral class B2V and earlier, whereas reflection nebulae are associated with stars of later spectral type.

Most associations of stars with large amounts of gas are related to either the birth or death of a star. Stars form by condensation in large complexes of gas, and the later stages of stellar evolution are usually characterized by high mass loss. Nebulae which are ionized by the young stars which formed out of the gas cloud are called HII regions or Strömgren spheres. Nebulae which owe their existence to the ejection of gas associated with the final stages of stellar evolution are called either planetary nebulae (if the ejection is mild), or novae/supernovae remnants (if the ejection is violent).

Giant HII regions, of which the Orion Nebula (M42) is an outstanding example, are among the most luminous objects in the Galaxy. Because of the tendency for young O stars to form in associations, the larger HII regions are usually ionized by a number of stars. Characteristically, HII regions are very large, massive complexes of gas having relatively low density in comparison with the smaller, less massive, but more dense, planetary nebulae. The latter objects are the result of a moderately non-violent form of mass loss by stars which are generally believed to have originated near the blue end of the horizontal branch, and are the precursors of white dwarfs. The central stars of planetaries are extremely hot, condensed objects. The general properties of HII regions and planetary nebulae, and their exciting stars, are summarized in Table 3.1. There are no strict definitions for the various kinds of nebulae, and there occasionally appear in the literature debates over the classification of certain objects. The wise should be guided by Minkowski's (1968) suggestion that a planetary nebula is "an object that appears in a catalogue of planetary nebulae, and to which there are no serious objections."!

The present discussion concerns the physical processes that occur in gaseous nebulae as opposed to a study of the objects themselves. Unfortunately, the survey nature of this theoretical course forbids us from considering the observational aspect of the subject, which is very interesting in itself, and extremely useful in providing basic data concerning star formation, the dynamics of the interstellar medium, the abundances of the elements, galactic dy-

TABLE 3.1

General Properties of HII Regions and Planetary Nebulae

Characteristic	HII Regions	Planetary Nebulae
Size	10 pc	0.1 pc
Density	1 - 100 cm^{-3}	~ $10^4 cm^{-3}$
Mass	$\gtrsim$ 100 $M_\odot$	1/2 $M_\odot$
Temperature of Ionizing Star	~40,000°K	100,000°k
Lifetimes	10^7 years	10^4 years
Spatial Distribution in Galaxy	Pop I objects, confined to plane	Intermediate disk Pop II objects
Relationship to Ionizing Star	Ionizing stars form from the gas	Nebula is ejected by an evolved star

namics, and later stages of stellar evolution. Instead, we will focus our attention on the general problem of deducing the physical conditions in nebulae, with the understanding that the discussion has application not only to nebulae, but also to any objects which display a similar emission-line spectrum, e.g., the nuclei of Seyfert galaxies, and quasars.

Thermodynamic Equilibrium vs. Steady-State Equilibrium. In Chapter 1, the distinction was drawn between the concept of local thermodynamic equilibrium (LTE) and statistical equilibrium. In order to determine the applicability of LTE to gaseous nebulae, it is instructive to consider the changing conditions in the gas as one goes out from the center of a star through the atmosphere, and into the nebula. At each point in the interior of a star, where LTE exists, there is a unique temperature which describes the energy distribution of photons and particles. As noted in Chapter 2, the excitation and ionization of the gas in this situation are prescribed by the Boltzmann and Saha equations, and the radiation field is characterized by the Planck function. Recall that there are two conditions that must be fulfilled if LTE is to hold for a point in a gas. First of all, the radiation field and the matter must be strongly coupled to each other (through emission and absorption processes), such that any changes in the characteristics of one can be immediately and thoroughly communicated to the other. Secondly, the mean free paths of both particles and photons must be less than the distance over which the temperature undergoes any appreciable change. Thus LTE requires a strong interaction between radiation and matter, and relative isolation from regions of the gas where the temperature is different. Since reaction rates always increase with the densities of the interacting particles, and the mean free paths of both particles and photons decrease with increasing density, both the matter-radiation coupling and the isolation of a system are greater at higher densities. Consequently, it is a necessary (and sufficient) condition for LTE to apply in a gas that the particle and radiation densities be very high. This situation occurs in the interiors of stars.

As one goes out from the center of a star, the condition of hydrostatic equilibrium dictates a decrease in the gas density. Furthermore, the fact that nuclear energy generation occurs almost exclusively near the star's center leads, from conservation of energy, to a decrease in the energy density of radiation with increasing distance from the center. Consequently, interactions between the radiation field and matter occur less frequently, and there is increased "contamination" of local interactions by particles and photons which originate at distant points in the star, which are characterized by different temperatures. As a result, by the time the outer layers of the star are reached, LTE no longer strictly applies, although it can still be used to give a rough approximation of actual conditions in the atmosphere (cf. Chapter 1).

The dimensions of most nebulae are orders of magnitude larger than the sizes of the stars they surround, so in nebulae the energy density of the radiation field, which originates in the star, is very low. In addition, the typical nebular gas density is much lower than that found anywhere in a star. Thus, in this type of environment conditions differ greatly from those required to insure LTE, and therefore descriptions of the ionization, level populations, etc., must be obtained from some other assumption than LTE. The assumption usually made for gaseous nebulae is that they are in a steady-state, or quasi-static, equilibrium called statistical equilibrium. That is, all of those processes which are important in determining the equilibrium value of a quantity are assumed to occur at a rate which is much less than the time-scale of the evolution of the gas. As an example, we can illustrate this point easily by applying it to the excitation conditions in a nebula.

Excitation Conditions in a Dilute Radiation Field. Consider an ensemble of 2-level atoms which are excited by black-body radiation from a star, having temperature T_* and radius R_*, whose distance is r. The mean intensity of radiation $J_\nu = W B_\nu (T_*)$, where $W = R_*^2/4r^2$ is the geometrical dilution factor, and $B_\nu (T_*)$ is the Planck function. When $W << 1$, the radiation field is said to be "dilute". In a steady-state situation, the number of atoms in each level remains constant in time. Therefore, the transition rate of $1 \rightarrow 2$ excitations per unit time must equal the rate of $2 \rightarrow 1$ de-excitations per unit time. If we consider only radiative transitions between the levels to be important, then by definition of the Einstein A and B coefficients, the

$$\text{No. of excitations } 1 \rightarrow 2/(\text{cm}^3 \text{ sec}) = N_1 B_{12} J_{\nu_{12}}, \qquad (3.1)$$

$$\text{No. of de-excitations } 2 \rightarrow 1/(\text{cm}^3 \text{sec}) = N_2 (A_{21} + B_{21} J_{\nu_{21}}). \qquad (3.2)$$

The steady-state condition for the relative population of the two levels comes from equating these two rate equations:

$$\frac{N_2}{N_1} = \frac{B_{12} J_{\nu 21}}{A_{21} + B_{21} J_{\nu_{21}}}, \qquad (3.3)$$

where $J_{\nu_{12}} = J_{\nu_{21}}$ is the mean intensity at the line frequency ν_{21}. Using the relations derived in Chapter 1 between the Einstein coefficients, $A_{21} = B_{21} \; 2h\nu^3/c^2$ and $g_1 B_{12} = g_2 \; B_{21}$, and the Wien approximation, $B_{\nu_{21}}(T_*) = 2h\nu_{21}^3/c^2 \; e^{-h\nu_{21}/kT_*}$, it follows that

$$\frac{N_2}{N_1} = W \frac{g_2}{g_1} \; e^{-h\nu_{21}/kT_*}, \qquad (3.4)$$

a result which differs from Boltzmann's equation by the presence of the dilution factor. For distances much greater than a stellar radius, $W \ll 1$, and one has the result that level 2 is very underpopulated with respect to level 1 in comparison with its population in LTE. The reason for this result lies in the fact that the rate for $2 \rightarrow 1$ transitions per ion $\sim A_{21}$ when $W \ll 1$, independent of conditions in the gas. That is, the probability that an atom spontaneously de-excites by radiating away energy is a physical constant. On the other hand, the rate for a $1 \rightarrow 2$ excitation $\sim J_{\nu_{12}}$ because it requires the absorption of a photon. Consequently, the excitation rate is drastically reduced in a dilute radiation field. Although we have not taken <u>collisional</u> excitation and de-excitation processes into account in the derivation of this result, it is still generally valid in nebulae: in the highly dilute radiation field typical of gaseous nebulae, the vast majority of all atoms and ions occupy their ground states.

3.2 Important Atomic Processes in Gaseous Nebulae

Because the normal relationships valid for LTE do not hold in gaseous nebulae, physical conditions within the gas must be determined by considering a detailed analysis of all relevant interaction processes. As explained in Chapter 1, this type of analysis is called a "microscopic" analysis because it requires a knowledge of each individual process, as opposed to the "macroscopic" analysis that can be used for a gas in LTE, which is based upon relationships derived from general thermodynamic principles.

All of the important interactions that occur in nebulae can be classified either as particle-photon reactions or as particle-particle encounters. We will consider what some of these reactions are, and then derive general expressions for the corresponding reaction rates. The following interactions have application to gaseous nebulae:

(1) Photoionization and Excitation - a photon interacts with an atom or ion, causing ejection or excitation of a bound electron. The photon is absorbed. There are conditions upon the energy of the incident photon for both of these reactions to occur. Photoionization requires the energy of the photon to exceed the ionization potential of the ion, $h\nu \geq \chi_I$. Any excess energy is carried off by the photoelectron as kinetic energy. Photoexcitation requires the photon energy to be the same as the difference in energy between the initial and final levels occupied by the bound electron.

(2) Collisional Ionization and Excitation - a free electron interacts with an ion or atom, causing ejection or excitation of a bound electron. The free electron is not captured in this process, unlike the photon in photoionization, because the electron can give up a fractional part of its free energy. Otherwise, processes (1) and (2) are similar. The threshold required for collisional ionization or excitation to occur is that the kinetic energy of the free electron exceed the ionization or excitation potential, $1/2\, mv^2 \geq \chi$.

Collisional de-excitation, the inverse process to collisional excitation, is occasionally an important process for certain ions. It occurs when a free electron interacts with the excited bound electron of an ion, and the bound electron makes a transition to a lower level, giving additional kinetic energy to the free electron. Collisional de-excitation is a similar, but competing, process to spontaneous radiative de-excitation. In both cases, a bound electron de-excites, decreasing its potential energy. In radiative de-excitation, the energy goes into the creation of a photon, whereas in collisional de-excitation the energy is simply given to the free electron.

(3) Electron Recapture (Recombination) - a free electron is recaptured by a positive ion, causing a decrease in its ionization. The excess kinetic energy of the electron is radiated away (free-bound continuum emission).

(4) Electron-Electron Collisions - free electrons have Coulomb encounters with each other which serve to redistribute the velocities of each electron. The collisions are essentially elastic, so no radiation results.

Particle-Particle Collision Rates. A general expression can be derived for the rate at which collisions between particles occur, which depends upon the interaction cross section. Consider a system of two kinds of particles: A and B. Define a cross section for interactions between A and B particles $\sigma = \pi r_o^2$, which is defined in such a way that, statistically, the interaction occurs if the distance between any 2 particles $\leq r_o$. Let N_A and N_B represent the number density (cm^{-3}) of the particles, and suppose that the A particles are at rest, whereas the B particles have a speed distribution function $f(v)dv$. Then, the

$$\text{No. of Bs/(cm}^3\text{) having speed } v \text{ to } v+dv = N_B f(v)dv. \quad (3.5)$$

In a time interval dt, one B particle with velocity v sweeps out an "interaction volume" $\sigma v dt$, interacting with all A particles within the volume. Hence, the

$$\text{No. of interactions between one B particle, of velocity } v \text{, and all As/sec} = N_A \sigma v dt/dt. \quad (3.6)$$

Since there are $N_B f(v)dv$ particles/cm^3 having velocities between v and v + dv, the total number of interactions between the A and B particles per unit volume and time is obtained by multiplying the above two equations and integrating over all velocities, i.e., the

$$\text{No. of AB interactions}/(\text{cm}^3\ \text{sec}) = N_A N_B \langle\sigma v\rangle, \qquad (3.7)$$

where $\langle\sigma v\rangle \equiv \int_0^\infty \sigma v f(v) dv$, and σ is usually velocity-dependent. This expression is a general one that can be used to determine the interaction rate for any type of collision once the cross section has been calculated or measured in the laboratory.

Radiative Interaction Rates. The rate for particle-photon interaction processes is governed by the equation of transfer, which gives the change in intensity of radiation dI_ν along a path length ds,

$$\frac{dI_\nu}{ds} = -\kappa_\nu I_\nu + j_\nu. \qquad (3.8)$$

Confining our attention to absorption processes only ($j_\nu = 0$), we can write that the

$$\text{Energy absorbed}/(\text{cm}^3\text{sec Hz ster}) \equiv \left|\frac{dI_\nu}{ds}\right| = \kappa_\nu I_\nu = N\ a_\nu I_\nu, \qquad (3.9)$$

where N is the number of absorbers/cm^3, and a_ν is the absorption cross section per absorber (cm^2). To obtain the number of absorptions, i.e., interactions, per unit volume and time, the above expression must be divided by the energy of each photon, $h\nu$, and integrated over all frequencies and directions,

$$\text{No. of absorptions}/(\text{cm}^3\ \text{sec}) = N\int_0^\infty \frac{4\pi J_\nu a_\nu}{h\nu}\ d\nu, \qquad (3.10)$$

since $J_\nu = \frac{1}{4\pi} \int_{4\pi} I_\nu d\omega$. This expression is analogous to the one derived for particle-particle collisions, and, in fact, can be written in the same form as equation (3.7) if one deals with the number density of photons ($N_\nu = u_\nu/h\nu = 4\pi J_\nu/ch\nu$), rather than the mean intensity of radiation.

Cross Sections. Since conditions in gaseous nebulae are very difficult to duplicate in a laboratory, most of the cross sections required for the study of nebulae must be obtained from theory. This general problem is beyond the scope of the present discussion, and we will simply quote a few relevant results which we shall use. As a general example of how a rough cross section may be computed, however, we will consider the cross section for electron-electron collisions. Since Coulomb repulsion acts over an infinite distance, we must first define what an e-e encounter is. The only effect of e-e collisions is a re-distribution of velocities. Therefore, we arbitrarily define the cross section, σ_e, for e-e scattering in terms of the distance r_o at which the electrostatic energy of two electrons is comparable to their relative kinetic energy. That is, $\sigma_e = \pi r_o^2$ where

$$\frac{e^2}{r_o} \sim 1/2\, mv^2 \,, \tag{3.11}$$

$$\therefore \qquad \sigma_e = \frac{4\pi e^4}{m^2 v^4} \,. \tag{3.12}$$

If the electrons have a Maxwellian speed distribution function, then the r.m.s. velocity is given by $1/2m\langle v^2\rangle = 3/2kT_e$, and so

$$\sigma_e(\langle v^2\rangle^{1/2}) = \frac{4\pi e^4}{9(kT_e)^2} \sim 4 \times 10^{-14}\text{cm}^2 \quad \text{for } T_e = 10^4 {}^\circ\text{K}. \tag{3.13}$$

This expression has the same functional dependence upon velocity as a rigorous derivation, and has essentially the correct numerical value.

The photoionization cross section for the nth level of hydrogen is

$$a_n(\nu) \simeq 7 \times 10^{-18}\, n\left(\frac{\nu_n}{\nu}\right)^3 \text{ cm}^2 \qquad \text{for } \nu \geq \nu_n, \qquad (3.14)$$

$$= 0 \qquad \text{for } \nu < \nu_n,$$

where ν_n is the ionization frequency of the nth level.

The recombination cross section for the capture of a free electron to the nth level of H is

$$\sigma_n^{rec}(v) \simeq \frac{\sigma_o}{n}\left(\frac{v_o}{v}\right)^2 \text{ cm}^2, \qquad (3.15)$$

where $\sigma_o \simeq 2 \times 10^{-21} \text{cm}^2$, and $v_o \simeq 5.5 \times 10^7$ cm/sec.

3.3 The Ionization Equilibrium

Nebulae have no energy source of their own. All their energy comes from the stars with which they are associated. Energy transfer is accomplished either by absorption of the star's radiation or the collision of outflowing stellar material with the nebula. In most situations the energy flux of stellar photons greatly exceeds the energy associated with stellar winds, so the basic energetics of nebulae is determined by the absorption of stellar radiation within the gas. Generally, nebulae have roughly cosmic, or solar, abundances [N(He)/N(H) ~ 10^{-1}, N(C, N, O, Ne)/N(H) ~10^{-4}], therefore H is the most important source of opacity in the gas, by virtue of its larger abundance. In its neutral state, hydrogen (HI) is virtually all in the ground state, so it absorbs both Lyman line and continuum (ionizing) photons. Ionized hydrogen (HII), on the other hand, is a very ineffective absorber of radiation. Because there are much fewer photons radiated by most stars in the Lyman lines than in the ionizing continuum (λ < 912Å), photoionization of hydrogen is the primary process through which nebulae receive energy.

The basic energy cycle by which gaseous nebulae convert stellar far-ultraviolet radiation to other wavelengths is as follows. In its lowest energy configuration, hyrdogen is neutral and in the ground state. Radiation incident upon the gas having frequencies of the Lyman lines or continuum excites and ionizes the hydrogen. Radiation of all other frequencies passes through the gas without any interaction, e.g., visible light is unattenuated by a nebula unless a substantial amount of dust is present. Photons with Lyman line wavelengths, $\lambda\lambda$1216, 1026, etc., will excite neutral H to excited levels, after which the atom returns to the ground state, either directly or via intermediate levels by spontaneous emission. The net effect of the Lyman line absorption is the conversion of a Lyman line photon to other line photons, say, a Paschen or Balmer line photon plus a different Lyman line ($L\gamma \rightarrow P\alpha + L\beta$, or $H\beta + L\alpha$). However, this process is of little consequence to the gas.

Photons in the Lyman continuum ionize the neutral H, creating a free proton-free electron pair. The ejected photoelectron carries away kinetic energy, which it gives to other particles before its ultimate recapture by a proton. The two most common interactions suffered by free electrons before recapture are (1) Coulomb collisions with other free electrons, a process having a very large cross section in comparison with other competing processes, and (2) collisional excitation of ions of the heavy elements. The first process causes the electrons to have a Maxwellian velocity distribution, while the second reaction serves as an energy sink for the electrons. After collisional excitation by a free electron, an ion usually de-excites radiatively, so the kinetic energy of the electrons goes directly into line radiation from the heavy ions in the gas. Eventually, the electrons are recaptured by protons — a process which creates free-bound continuum radiation. Most recombinations are to the excited states, whereupon the atoms spontaneously radiatively de-excite down to the ground state. In doing so, they create line radiation which is observed from nebulae, including the Balmer lines ($H\alpha$, $H\beta$, ...) and radio lines produced by transitions between states of very high quantum numbers. This cycle of ionization by stellar radiation, followed by eventual recombination is repeated over and over again. The net result of the cycle is a conversion of the stellar UV radiation to nebular emission lines and continuum.

The fundamental problem in the study of nebulae is the determination of the state of ionization of the gas. Saha's equation cannot be used because LTE does not hold in the gas. Instead, it must be assumed that the nebula is in a steady-state, and the ionization is not changing in time. Then, at each point in the gas, the rate at which H (or any other element) is photoionized must equal the rate at which recombinations occur, since these are the only two processes of any importance which involve changes in the ionization of hydrogen. According to equations (3.7) and (3.10), this condition requires that

$$N_{HI} \int_{\nu_1}^{\infty} \frac{4\pi J_\nu a_1(\nu)}{h\nu} \, d\nu = N_e N_{HII} \sum_{n=1}^{\infty} \langle \sigma_n^{rec} v \rangle , \qquad (3.16)$$

where J_ν is the mean intensity of ionizing radiation ($\nu>\nu_1$), $a_1(\nu)$ is the ground state photoionization cross section of H, and σ_n^{rec} is the recombination cross section to the nth level of hydrogen. A similar ionization equation applies to each ion of every element in the gas, as long as photoionization and recombination are the primary processes governing ionization.

In order to put the ionization equation in a more viable form, the principle of detailed balancing can be used to derive a relationship between $a_n(\nu)$ and $\sigma_n^{rec}(v)$ (Milne 1924), in the same way that detailed balancing was used in Chapter 1 to derive the relation between the collisional excitation and de-excitation coefficients. If the ionizing star is assumed to radiate like a black body, then it can be shown that the ionization equation takes on a form similar to the Saha equation,

$$\frac{N_{HII} N_e}{N_{HI}} = \left(\frac{2\pi m_e k T_*}{h^2}\right)^{3/2} e^{-\chi_H/kT_*} \sqrt{T_*/T_e} \; W e^{-\tau_\nu} , \qquad (3.17)$$

where χ_H is the ionization potential of H, T_e is the kinetic temperature of the free electrons, and τ_ν is the optical depth into the gas from the star. This result is due to Strömgren (1948). Interestingly enough, this equation was actually <u>solved</u> for the ionization in an earlier paper by

Strömgren (1939), almost 10 years before it was really derived. An approximate form of the equation had been heuristically deduced from Saha's equation by Rosseland (1936), and was used by Strömgren in his initial paper, which was the first theoretical investigation concerning the ionization of the interstellar medium. It wasn't until much later that Strömgren rigorously derived the correct ionization equation.

Unfortunately, the ionization equation does not lend itself to a straightforward analytical solution. The optical depth τ_ν involves an integral of the density of neutral H, N_{HI}, through the gas, consequently equation (3.17) is a complicated integral equation. We will only describe the qualitative aspects of the solution; a detailed treatment may be found in Strömgren (1939) and Seaton (1960).

Hydrogen has only two stages of ionization; it is either neutral or ionized. Near a hot star, the mean intensity of ionizing radiation is relatively large, since $W \sim 1$, and $\tau_\nu \sim 0$. In addition, $(2\pi m_e kT_*/h^2)^{3/2} >> 1$, and the electron density N_e is relatively low. Thus, $N_{HII}/N_{HI} >> 1$ surrounding a hot star, and the hydrogen is therefore fully ionized. Two factors eventually cause the ionization to decrease with increasing distance: (1) the $1/r^2$ geometrical diminution of the radiation, and (2) the depletion of ionizing photons by absorption, i.e., increasing optical depth. Because of the much greater sensitivity of the radiation field to optical depth than distance, the ionization remains high until $\tau_\nu \sim 10$ at frequencies where most of the ionizing photons occur (usually near ν_1). When $\tau_{\nu_1} \sim 10$ the stellar ionizing photons are virtually all absorbed, so the ionization drops, and the gas becomes neutral. That is, near the star there is a large supply of ionizing photons which are capable of sustaining ionization. In causing ionization, however, the photons are depleted through absorption. Further out into the gas, the ionizing flux gets smaller and smaller until a point is reached where the radiation field is completely absorbed, and the gas then becomes neutral. The characteristic distance over which the radiation field drops from a value capable of maintaining ionization to the point where it is essentially completely absorbed is roughly one mean free path for a photon. At the point where the ionization is beginning to drop, this distance is usually much smaller than the characteristic size of the ionized region, and therefore the demarcation between the ionized (HII) region and the neutral (HI) region is usually very sharp.

Sizes of HII Regions. The extent of an HII region can be determined by solving the ionization equation, above. However, there is a more straightforward manner of calculating the size of an HII region: apply the steady-state condition for ionization to the nebula as a whole. Then, the total number of ionizations in the nebula/sec must equal the total number of recombinations in the nebula/sec. Or, since every UV ($\lambda < 912Å$) photon emitted by the star produces an ionization,

$$\int_{\nu_1}^{\infty} \frac{L_\nu}{h\nu} d\nu = N_e N_{HII} \sum_{n=2}^{\infty} <\sigma_n^{rec} v> 4/3\ \pi R_{HII}^3. \quad (3.18)$$

The reason recombinations to excited states only are considered in the above equilibrium is that recombinations to the ground state create free-bound ionizing photons which are simply re-absorbed by the gas, so these two processes occur at equal rates, and therefore cancel each other. If the star is assumed to have a Planckian flux distribution, then

$$\int_{\nu_1}^{\infty} \frac{L_\nu}{h\nu} d\nu = \frac{8\pi^2 R_*^2}{c^2} \left(\frac{kT_*}{h}\right)^3 F_1(x_1), \quad (3.19)$$

where R_* and T_* are the stellar radius and temperature, $x_1 \equiv h\nu_1/kT_*$, and $F_1(x_1)$ is defined as

$$F_1(x_1) = \int_{x_1}^{\infty} \frac{x^2}{e^x - 1} dx. \quad (3.20)$$

The function $F_1(x_1)$ has been tabulated by Zanstra (1931) and Harman and Seaton (1966), and is evaluated in Table 3.2 for several values of x_1 by numerical integration.

Assuming that $N_e = N_{HII} = N$ in the HII region, where N is the gas density (cm^{-3}), and making use of the fact that $\sum_{n=2}^{\infty} <\sigma_n^{rec} v> = 2.6 \times 10^{-13}\ cm^3 sec^{-1}$ for an electron tem-

TABLE 3.2

The Integral $F_1(x_1)$

T_*(°K)	x_1	$F_1(x_1)$
10,000	15.79	3.91×10^{-5}
15,000	10.53	3.58×10^{-3}
20,000	7.90	2.98×10^{-2}
25,000	6.32	9.85×10^{-2}
30,000	5.26	0.209
35,000	4.51	0.346
40,000	3.95	0.495
45,000	3.51	0.645
50,000	3.16	0.790
60,000	2.63	1.05
70,000	2.26	1.26
80,000	1.97	1.43
90,000	1.75	1.57
100,000	1.58	1.69

perature typical of nebulae, $T_e = 10^4$°K, equation (3.18) can be solved to give

$$R_{HII} = 78\ N^{-2/3} \left(\frac{R_*}{R_\odot}\right)^{2/3} \frac{F_1^{1/3}(x_1)}{x_1}\ \text{pc}. \qquad (3.21)$$

This expression generally gives overestimates of the actual size of HII regions around stars of given effective temperature T_*, because of the assumption of a Planckian radiation field for the star. Model atmosphere calculations for hot stars (cf. Mihalas 1972, and references therein) show that there is usually a deficiency in the number of stellar photons emitted below the Lyman limit in comparison with the radiation emitted by a black body of the stellar effective temperature. The decrease in flux for $\lambda <$ 912Å is due to

TABLE 3.3

HII Region Sizes for Different Stars (N=1 cm^{-3})

T_*(°K)	$R_*/R_\odot$	Spectral type	Flux Distribution	T_{eff} (λ < 912Å)	R_{HII}(pc)
50,000	15.1	O4 V	Black body	50,000	135
			Model atmos.	51,300	137
45,000	12.6	O5.5 V	Black body	45,000	101
			Model atmos.	45,800	101
40,000	10.2	O6.5 V	Black body	40,000	72
			Model atmos.	39,700	69
35,000	7.9	O9 V	Black body	35,000	48
			Model atmos.	32,300	40
30,000	7.6	BO V	Black body	30,000	35
			Model atmos.	21,900	16
22,500	6.2	B1 V	Black body	22,500	15
			Model atmos.	12,800	2.2
20,000	5.6	B2 V	Black body	20,000	10.3
			Model atmos.	11,500	1.3
17,500	4.4	B3 V	Black body	17,500	5.8
			Model atmos.	10,300	0.6

HI continuum absorption in the photosphere. This effect becomes smaller at higher stellar temperatures because of the increasing ionization of the photospheric hydrogen. In order to demonstrate the importance of this departure from a Planckian spectrum on the sizes of HII regions, we have computed the run of ionization of hydrogen out from several stars using the model atmospheres of Mihalas (1972) and the stellar data compiled by Panagia (1973), and the results are given in Table 3.3. We have listed the effective (bolometric) temperature T_* and corresponding spectral type of the ionizing stars, and the computed HII region size for a star of that temperature if it were to radiate like a black body. Also listed is a determination of the effective temperature of each of the stars below λ912Å on the basis of Mihalas' computations, i.e., the equivalent temperature a

TABLE 3.4

Ionization Potentials for Ions of Abundant Elements

Ion	Ionization Potential (eV)	λ_1 (Å)	Ion	Ionization Potential (eV)	λ_1 (Å)
MgI	7.644	1622	HeII	54.40	228
CI	11.264	1101	OIII	54.89	226
HI	13.595	912	NeIII	63.5	195
OI	13.614	911	CIV	64.48	192
NI	14.53	853	OIV	77.39	160
MgII	15.03	825	NIV	77.45	160
NeI	21.56	575	MgIII	80.12	155
CII	24.38	509	NeIV	97.02	128
HeI	24.58	504	NV	97.86	127
NII	29.59	419	MgIV	109.3	113
OII	35.11	353	OV	113.9	109
NeII	41.07	302	NeV	126.3	98
NIII	47.43	261	OVI	138.1	90
CIII	47.86	259	MgV	141.2	88

black body would have in order to emit the same total flux below the Lyman limit as the model atmosphere, and the size of the computed ionized region of gas. All calculations have been made assuming a constant gas density of $N = 1 cm^{-3}$. It is apparent that the deviations of stellar fluxes from a Planckian distribution can produce large differences in the sizes of the nebulae. It is also apparent that HII region sizes are very sensitive to the spectral type of the exciting star, because of the roughly exponential dependence of the number of emitted ionizing photons upon the stellar temperature.

Ionization of the Heavier Elements. Photoionization and recombination cross sections tend not to vary greatly from one ion or element to another. Thus, the most important factor governing the ionization of an element is the number of photons available that are capable of producing ionization. The total ionizing flux of radiation for an ion depends sensitively upon the ionization potential of the ion. Most elements of high cosmic abundance have ionization potentials greater than that of hydrogen (cf. Table 3.4). H and He are much more abundant than any of the other elements, so they are the only significant sources of continuous opacity in a nebula. HI absorbs photons of $\lambda \leq 912Å$, HeI absorbs radiation of $\lambda \leq 504Å$, and HeII absorbs radiation of $\lambda \leq 228Å$. Because of the tendency for photoionization cross sections to be largest near threshold frequencies, these ions preferentially absorb radiation at wavelengths immediately shortward of their ionization limits. This leads to an ionization structure of the gas which can be understood in terms of distinct H and He ionization zones.

In general, an element will be ionized by stellar radiation in a nebula as long as there are unabsorbed ionizing photons available for ionization. The ionization process serves as a sink for such photons, and when they are all depleted by photoionization, the ionization drops. In most astrophysical situations, the flux of radiation below 912Å decreases with decreasing wavelength, as, for example, occurs with Planckian radiation at temperatures below 10^5°K, and synchrotron radiation. Therefore, there are usually fewer photons capable of ionizing HeII than HI, because of the much higher ionization potential of HeII. Normally, the ionization in a nebula is highest near the source of radiation, where the radiation field is most intense, and it decreases outward. Initially, there will be a region next to the star in which the hydrogen and helium are completely ionized: HII, and HeIII. The heavy elements in this region reside in those stages of ionization which have ionization potentials greater than that of HeII (228Å): CV CVI, NV, OIV, OV, NeIV, etc. Because there are few HeII ionizing photons than HI or HeI ionizing photons, the radiation field below 228Å becomes depleted by absorption from HeII → HeIII ionization long before the HI ionizing photons (λ < 912Å) are absorbed, and a distance is eventually reached where the helium become HeII. There is essentially no radiation shortward of λ228Å outside of the

HeIII ionization zone, so no ion can exist with an ionization potential greater than that of HeII (54 eV) in the region where the He is HeII. Hence, the abundant heavy ions in the HeII and HII region are CIV, NIV, OIII, NeIII, etc. As one proceeds further from the ionizing star, the ionization of HeI → HeII and HI → HII depletes the radiation below λ504Å and λ912Å, and the helium and hydrogen eventually become neutral. Generally, these two elements go neutral at about the same distance. This occurs because HI, by virtue of its greater abundance, absorbs more λ <504Å radiation than does HeI. Furthermore, since the photoionization cross section of H is greatest near 912Å, photons of wavelengths λ < 912Å are absorbed before photons of λ <504Å. Thus, ionization of H and He in the periphery of a nebula is sustained by radiation below the <u>HeI</u> ionization limit. When all these photons are absorbed, the H and He become neutral. At the same distance, all other elements also go neutral except those having ionization potentials less than that of H. C, Mg, Na, Si, and Ca are among the cosmically more abundant elements which remain singly ionized when H is neutral. Figure 3.1 shows a typical ionization distribution for several elements in an HII region. The fractional abundance x_i of each ion has been computed for a model nebula and is given as a function of distance from the exciting star, which was taken to be an O5.5V star ($T_* = 45{,}000$°K, $R_* = 12.6\ R_\odot$), assuming the gas density to be constant, $N = 1\ \mathrm{cm}^{-3}$. The far-UV flux from the star was taken from a non-LTE model atmosphere computed by Mihalas (1972), which shows deviations from black body radiation.

3.4 Kinetic Temperature and Energy Balance

The elastic Coulomb scattering of free electrons in electron-electron collisions occurs at a rate which exceeds almost all other particle collisions in an ionized gas because of its relatively large cross section (equation 3.13). These collisions serve to randomly distribute the velocities of the electrons, and therefore the distribution of electron velocities in an HII region is described by the Maxwell-Boltzmann function in terms of a kinetic temperature T_e, called the electron temperature,

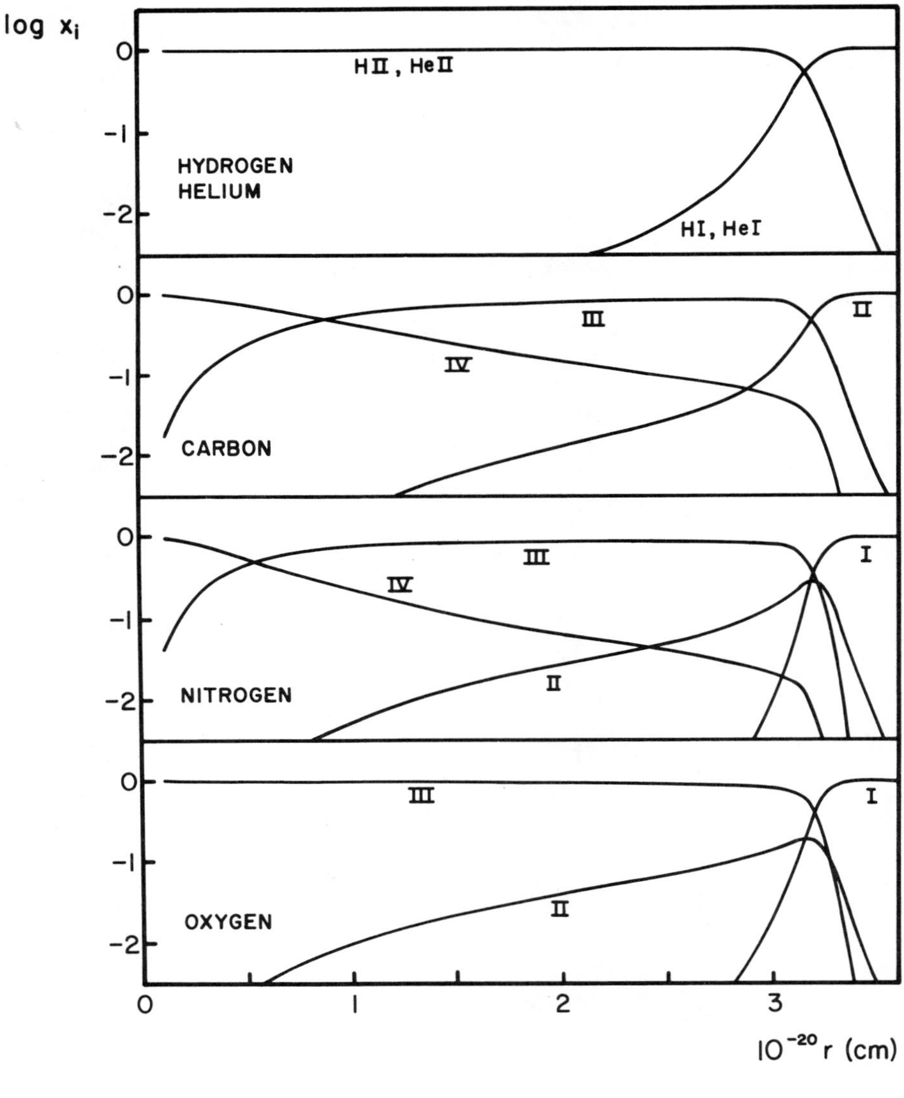

Fig. 3.1 The relative abundances x_i of the ions of each of the elements H, He, C, N, and O as a function of distance from an O5.5 V star, assuming the gas density to be $N=1\ cm^{-3}$

$$f(v)\,dv = 4\pi\left(\frac{m_e}{2\pi kT_e}\right)^{3/2} v^2 e^{-\frac{m_e v^2}{2kT_e}}\,dv. \qquad (3.22)$$

The electron temperature is an important quantity to be determined in nebulae because it governs the rate at which all particle reactions occur.

The electron temperature is determined by an equilibrium between processes which give the electrons energy and those processes which extract energy from the electrons. In a gas which is in a steady-state, the temperature does not change with time. Since T_e is a measure of the kinetic energy of the electrons (the mean energy per particle $\langle E\rangle = 3/2kT_e$), if it remains constant, the energy of the electrons must also remain constant in time. That is, at each point in the gas, the

$$\begin{matrix}\text{Energy gained by free} \\ \text{electrons/(cm}^3\text{sec)}\end{matrix} = \begin{matrix}\text{Energy lost by free} \\ \text{electrons/(cm}^3\text{sec).}\end{matrix} \qquad (3.23)$$

Since both the energy gain and loss rates depend upon T_e, the solution of this equation, sometimes called the equation of thermal equilibrium, determines the value of T_e at each point in a nebula.

Heating of the Electrons. Of the various atomic processes important in gaseous nebulae which we have considered, those that increase the electron energy are photoionization and collisional de-excitation. Collisional de-excitation is usually not an important process at the low densities found in most nebulae, except for a few selected transitions, because most excited ions radiately de-excite before collisional de-excitation of a level can occur. Thus, the only important heating mechanism for electrons is photoionization. When a free electron is created by ionization, it is given a kinetic energy $\frac{1}{2} m_e v^2 = h(\nu - \nu_1)$, where ν_1 is the ionization threshold frequency. Since hydrogen is by far the most abundant element, we can to a first approximation ignore other elements, and write that the energy gained by electrons/(cm^3 sec), G, is just the number of ionizations/(cm^3 sec) times the mean energy given to the ejected photoelectrons,

$$G = N_{HI} \int_{\nu_1}^{\infty} \frac{4\pi J_\nu a_1(\nu)}{h\nu} d\nu \times h(<\nu> - \nu_1), \tag{3.24}$$

where $h<\nu>$, the mean energy of ionizing photons, is given by

$$h<\nu> = \int_{\nu_1}^{\infty} \frac{J_\nu a_1(\nu)}{h\nu} h\nu d\nu / \int_{\nu_1}^{\infty} \frac{J_\nu a_1(\nu)}{h\nu} d\nu. \tag{3.25}$$

It is fairly straightforward to show that if the mean intensity has the spectral distribution of Planckian radiation in the Wien approximation, $J_\nu \propto \nu^3 \exp(-h\nu / kT_*)$, then it follows from equation (3.25) that $h<\nu> \simeq h\nu_1 + kT_*$. That is, the mean energy of an ejected photoelectron is just kT_*. Since ionization equilibrium requires the photoionization and recombination rates to be equal, the heating rate may be re-written as

$$G \simeq N_e N_{HII} \sum_n <\sigma_n^{rec} v> kT_*. \tag{3.26}$$

For a fixed degree of ionization, a hotter star heats the electrons more effectively than a cool star because the mean energy of ionizing photons is greater. The dependence of G upon the electron temperature occurs through the recombination coefficient. The velocity dependence of the recombination cross section is $\sigma_n^{rec} \propto v^{-2}$, and therefore $<\sigma_n^{rec} v> \propto T_e^{-1/2}$ [cf. equations (3.7), (3.15), and (3.22)]. That is, the heating of the free electrons decreases as the temperature increases. This result can be understood in terms of the ionization equation (3.16). Within an HII region, where the ionization of H is essentially complete ($N_e \simeq N_{HII} \simeq N$), the amount of neutral H varies as the recombination coefficient, $N_{HI} \propto \sum_n <\sigma_n^{rec} v> \propto T_e^{-1/2}$. Consequently, when T_e increases, the recombination rate drops because of the increased energy of the electrons, and the amount of HI decreases. Since heating of the electrons occurs as a result of ionization of HI, the heating rate also decreases when T_e goes up.

Cooling of the Electrons. Two processes extract energy from the electrons: recombination, and collisional excitation. Electron recapture acts in just the opposite manner as photoionization; it eliminates the electron, and in doing so substracts an amount $1/2 m_e v^2$ from the energy of the free electrons for each recapture. This energy is returned to the radiation field by the creation of a free-bound photon. The energy loss rate per unit volume, L_r, for this process is simply the recombination rate times the mean energy given up by a recaptured electron. Since H is by far the most abundant element,

$$L_r = N_e N_{HII} \sum_n \langle \sigma_n^{rec} v \rangle \times 1/2\, m_e \langle v^2 \rangle. \quad (3.27)$$

Because the electrons have a Maxwellian velocity distribution, $\frac{1}{2} m_e \langle v^2 \rangle = \frac{3}{2} kT_e$. The resulting dependence of energy loss by recombination upon temperature is therefore $L_r \propto T_e^{1/2}$, assuming that the gas remains essentially fully ionized. That is, the loss rate increases as T_e increases, in spite of the decreasing rate of recapture, because of the greater increase in the mean energy per electron with higher temperature. It is clear from equations (3.26) and (3.27) that if electron recapture were the major source of cooling in a nebula, the equilibrium electron temperature would be $T_e \simeq T_*$. Actual temperatures in gaseous nebulae are lower than this value, however, because collisional excitation is almost always a more efficient source of cooling than recombination.

Collisional excitation occurs whenever a free electron interacts with an ion or atom, giving up kinetic energy to the excitation of a bound electron to a higher energy level. Conservation of energy requires that the difference in the energy of the levels $\chi = \frac{1}{2} m_e (v_i^2 - v_f^2)$, where v_i and v_f are the initial and final velocities of the free electron. There is obviously an energy threshold for this reaction, $\frac{1}{2} m v_i^2 \gtrsim \chi$, whereas there is no such threshold for electron recapture. On the other hand, a typical cross section for electron-ion excitation, $\langle\sigma^{exc}\rangle \sim 10^{-16} cm^2$, is far greater than the characteristic cross section for electron-proton recombination, $\langle\sigma^{rec}\rangle \sim 10^{-21} cm^2$. In a nebula, the free electron energy loss from collisional excitation of an ion i is given by the rate at which the interaction occurs multiplied by the energy given up by the electron per reaction χ, or

$$L_c = N_e N_i \langle\sigma^{exc} v\rangle \chi. \tag{3.28}$$

The velocity dependence of the excitation cross section for electron-ion encounters is $\sigma^{exc} \propto v^{-2}$, hence

$$\langle\sigma^{exc} v\rangle \propto \int_{v_\circ}^{\infty} v^{-2}\, v \frac{v^2}{T_e^{3/2}} e^{-\frac{m_e v^2}{2kT_e}} dv \propto T_e^{-1/2} e^{-\chi/kT_e}, \tag{3.29}$$

where $v_o = (2\chi/m_e)^{1/2}$ is the threshold velocity for excitation. Clearly, cooling by collisional excitation is very sensitive to the temperature of the electrons. If the mean energy of individual electrons is appreciably below the reaction threshold, $\frac{1}{2}m_e\langle v^2\rangle = \frac{3}{2}kT_e < \chi$, little cooling can occur. Conversely, when a significant fraction of free electrons have the threshold energy, collisional excitation can be an important energy sink because of the relatively large cross section for this reaction.

Exactly what collisional transitions serve as sources of cooling in nebulae depends upon the atomic structure of the individual ions. A priori, one might expect H and He to be important because of their high abundances. However, the excited levels of both H and He all lie more than 10 eV above the ground state, and therefore the electron temperature must well exceed 10^4 °K($kT \sim 1$ eV when $T = 10^4$ °K) before collisional excitation of either of these elements becomes appreciable. On the other hand, many ions of the heavy elements N, O, Ne, Mg, Si, S, etc. have excited levels within 2-3 eV above the ground state, and therefore these ions, in spite of their lower abundances with respect to hydrogen, are capable of cooling nebulae at temperatures lower than 10^4 °K. In fact, many heavy element ions have ground states which have fine-structure splitting of the individual levels, with energies of 0.01 - 0.1 eV, and excitation of these levels contributes to the cooling at re-

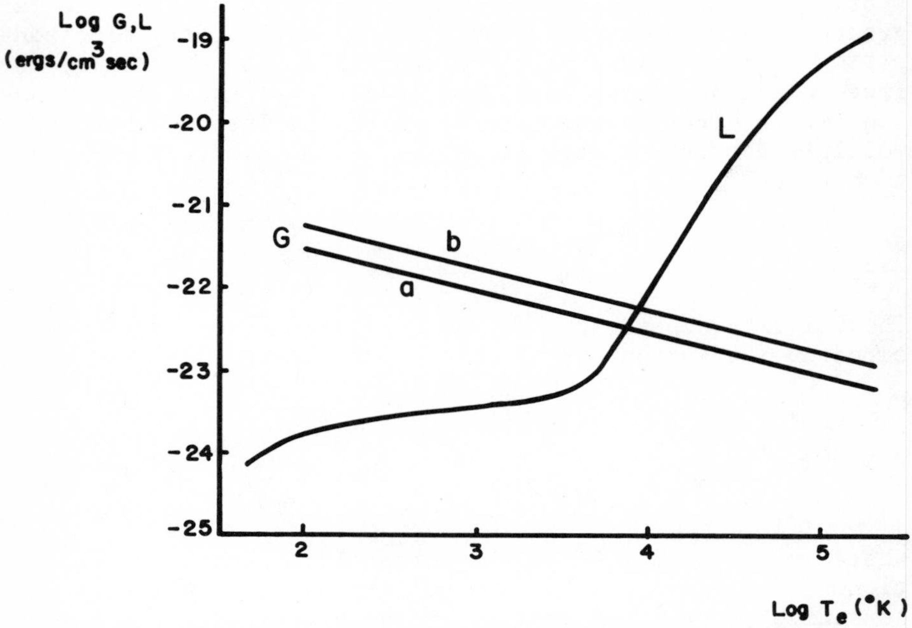

Fig. 3.2 Heating and cooling curves for the free electrons at a typical point in an HII region. The two heating curves correspond to ionizing stars of temperatures (a) 25,000 and (b) 50,000°K. The cooling is due almost entirely to collisional excitation of the heavy elements. Note that the equilibrium value of T_e is insensitive to characteristics of the ionizing star

latively low temperatures.

The total rate at which the electrons lose energy, $L = L_r + L_c$, has a complicated dependence upon the electron temperature. Collisional excitation of all the low-lying levels of the various ions of each of the heavy elements must be taken into consideration, in addition to the collisional excitation and recombination cooling of H and He. The cooling curve is fairly insensitive to the level of ionization in the gas because the ions of the heavy elements have similar energy levels over a broad range in ionization. For example, OII and NeV both have levels ~ 3 eV above the ground state. Consequently, the cooling of electrons does not change much from point to point in a typical nebula.

The equilibrium temperature of the electrons is set by the balance between heating and cooling processes, $G = L_r + L_c$ (equation 3.23). The solution of the thermal balance is depicted graphically in Figure 3.2. Curves representing both heating and cooling are given as a function of the electron temperature T_e . Solar abundances have been assumed for the element abundances. Heating has been assumed to be caused by black body radiation at temperatures of 25,000° and 50,000° K. The resulting equilibrium values of the electron temperature are 7250° and 8600°K, respectively. Because of the steepness of the cooling curve near log T_e~4, normal values of T_e in nebulae are generally around 10^4 °K, irrespective of conditions in the gas or characteristics of the exciting star. The sharp increase in the rate of cooling near temperatures of $T_e \sim 10^{4\circ}$ is caused by the onset of collisional excitation of many ions which have excited levels 2-3 eV above the ground state.

3.5 The Spectra of Gaseous Nebulae

A typical nebular spectrum consists of emission lines superimposed upon a faint continuum. The optical continuum is produced primarily by free-bound radiation resulting from electron recapture. The most prominent lines are the Balmer lines of hydrogen, Paschen α of HeII λ4686, and certain magnetic dipole transitions with low transition probabilities, called forbidden lines, of the heavy elements: [OII] λ3727, [OIII] λ5007, [NII] λ6584, [NeV] λ3426, [OI] λ6300, and [NeIII] λ3869. The emission-line spectrum can be understood in terms of the population of excited levels of ions and atoms, followed by radiative decay. All lines in the visible are optically thin, so all line photons which are created in the gas escape directly.

There are two basic mechanisms by which an excited state of an ion is populated: (1) recombination, and (2) collisional excitation from the ground state.

A. Recombination. Electrons are recaptured to excited states, and cascade downward, producing line radiation. Typically, recapture cross sections are $\langle\sigma_n^{rec}\rangle \sim 10^{-21}\,\mathrm{cm}^2$ for $T_e = 10^4$ °K. For a hydrogen-like ion, the recombination coefficient to the nth level is roughly

$$\langle\sigma_n^{rec} v\rangle \propto \frac{\text{const}}{n\,T_e^{1/2}} . \tag{3.30}$$

That is, electron recapture has a slight dependence upon the electron temperature, and is more likely to occur to lower levels. However, recombination does not have a threshold energy, and it is certainly not very sensitive to either n or T_e .

B. Collisional Excitation. Collisional excitation of an atom or ion virtually always occurs from the ground state (since almost all ions are in the ground state). Once excited, the ion usually radiatively de-excites. Collisional de-excitation does sometimes occur for certain transitions, in spite of the low nebular densities, and such situations will be discussed. Typical collisional excitation cross-sections at 10^4 °K are $<\sigma_n^{exc}> \sim 10^{-16}$ cm^2 , and there is a threshold for this reaction. The collisional excitation coefficient varies as (see equation 3.29)

$$<\sigma_n^{exc} v> \propto T_e^{-1/2} e^{-\chi_n/kT_e} , \tag{3.31}$$

so collisional excitation is very sensitive to both T_e and the energy of the excited level above ground state, χ_n.

Since $<\sigma^{exc}> >> <\sigma^{rec}>$ at nebular temperatures, the rate at which a level is populated by collisional excitation far exceeds that due to recombination, as long as $\chi_n \lesssim kT_e$. Therefore, at nebular temperatures of $T_e \sim 10^4$°K, all levels of an ion within about 5eV of the ground state are populated by collisional excitation. Higher levels are populated by recombination only. Since H and He have no excited levels within 10eV of the ground state, all excited levels of H and He are populated by recombination, and the observed emission lines of these two elements are called "recombination lines".

Unlike H and He, many ions of the heavy elements have excited levels within 2 - 3 eV of the ground state. These levels are populated by collisional excitation. Higher excited levels are populated by recombination, but at a much slower rate than the collisionally excited low-lying levels, because $<\sigma^{rec}> << <\sigma^{exc}>$. As a result, the most intense emission lines from the heavy elements come from the lowest levels. Heavy element recombination lines are seen in some nebulae, but they are generally very weak.

Transitions between the collisionally excited low-lying levels of the heavy elements are called "forbidden" transitions, because they are not normally observed under conditions found in terrestrial laboratories. They violate the electric dipole selection rules which apply to transitions (called "permitted") which have a high probability of occurring. There is a simple reason why these transitions tend to be forbidden. One of the electric dipole (permitted) selection rules is that transitions occur only between two levels in which the angular momentum quantum number ℓ changes, $\Delta\ell = \pm 1$. It so happens that the energies of atomic states are sensitive to ℓ. Therefore, any excited state having a different angular momentum than the ground state will tend to have a sufficiently higher energy such that it will not be collisionally excited. Conversely, excited states which are near enough to the ground state in energy to be collisionally excited at temperatures of 10^4 °K must have the same value of ℓ as the ground state. A radiative transition between the two levels must therefore be "forbidden". As we shall see, however, the fact that such transitions are forbidden has essentially no relevance to the intensities of these lines under normal conditions found in nebulae.

<u>Line Fluxes in Nebulae.</u> Much of what is known about gaseous nebulae has been deduced from the relative intensities of the emission lines. The line fluxes are dependent upon the abundances of the elements, gas density, electron temperature, and the ionization state of the gas, therefore each of these quantities can, in principle, be determined for a nebula from analysis of its spectrum.

It can easily be shown from the equation of transfer (cf. Chapter 1) that the expression for the emergent radiative flux F of an optically thin line emitted by an object having spherical symmetry with radius R, and characterized by a volume emission coefficient for the line j(r), is

$$F = \frac{4\pi}{R^2} \int_0^R j(r)\, r^2 dr. \qquad (3.32)$$

The emission coefficient for an emission line produced by a transition from level $n \rightarrow n'$ (neglecting stimulated emis-

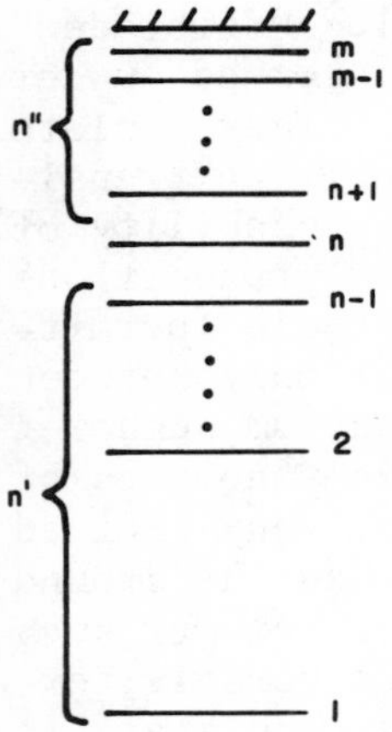

Fig. 3.3 A schematic energy level diagram for the excited levels of hydrogen, showing the nomenclature used in the text

sion, which is extremely unlikely in a dilute radiation field) is

$$j_{nn'} = N_n A_{nn'} \frac{h\nu_{nn'}}{4\pi} \qquad \text{ergs/(cm}^3\text{sec ster)}, \qquad (3.33)$$

where N_n is the number density of the upper level, $A_{nn'}$ is the Einstein coefficient, or transition probability, and $\nu_{nn'}$ is the line frequency. Since $A_{nn'}$ and $\nu_{nn'}$ are physical constants for the transition, the problem of determining line intensities simply reduces to the problem of establishing the population N_n of a given level. The level populations are determined by applying the steady-state condition to the processes causing population of a level. The two general methods by which excited states are populated are recombination from the continuum and collisional excitation from the ground state, and we shall consider each separately.

A. H and He Recombination Lines. The excited levels of both H and He are populated only by recombination, hence all line emission is a direct result of this process. Confining our attention to H, although the reasoning can be applied to He also, we will consider the problem of determining the emissivity of line radiation from transitions among the excited levels. Two related processes populate a given excited level of H: (1) direct electron recapture to the level, and (2) electron recapture to higher levels, followed by radiative cascading down into the level. De-population of excited states at nebular densities occurs only by radiative de-excitation. If we denote the various levels of hydrogen as designated in Figure 3.3, the equilibrium population of a level n is obtained by equating the rate at which the continuum and higher levels n'' populate level n to the rate at which it is depopulated by radiative transitions to lower levels, n'.

The rate of population of level n is given by the expression

$$\text{No. of atoms entering level } n/(\text{cm}^3\,\text{sec}) = N_e N_{HII} \langle \sigma_n^{rec} v \rangle + \sum_{n''=n+1}^{\infty} N_{n''} A_{n''n}, \quad (3.34)$$

where the first term represents direct recombination to n, and the second term represents radiative transitions into n from higher levels. The depopulation rate for level n is

$$\text{No. of atoms leaving level } n/(\text{cm}^3\,\text{sec}) = N_n \sum_{n'=1,2}^{n-1} A_{nn'}, \quad (3.35)$$

where the lower limit to the sum depends upon whether the nebula is optically thick or thin in the Lyman lines. If the nebula is optically thick in the Lyman lines, any transition to the ground state from a level n will produce emission of a Lyman-line photon, which will be immediately reabsorbed, causing re-population of the level. That is, detailed balancing holds between each excited state and the ground state in an optically thick gas, with the result that radiative de-excitation to the n=1 level does not produce any net de-population of excited levels. Since most nebulae are believed to be optically thick in the Lyman lines, this fact has interesting consequences for the population of the n=2 level of hydrogen. If every $2 \rightarrow 1$ (Ly-α) transition produces a photon which is reabsorbed, causing a $1 \rightarrow 2$ transition, there is no way of depopulating the first excited level, and its population will become very large, driving up the optical depth in the Balmer lines. What actually happens is that Ly-α photons eventually escape an optically thick gas, after they have been scattered many times.

If a nebula is evolving on a time-scale that is much longer than the time-scale in which excited levels are populated and de-populated, the populations N_n will achieve an equilibrium in which the rates of atoms entering and leaving any level n are equal, or

$$N_e N_{HII} \langle \sigma_n^{rec} v \rangle + \sum_{n''=n+1}^{\infty} N_{n''} A_{n''n} = N_n \sum_{n'=2}^{n-1} A_{nn'}. \quad (3.36)$$

This set of equations for the excited states of hydrogen is called the capture-cascade equations. The solutions of the equations for each level give the N_n as a function of the continuum population $N_e N_{HII}$. These equations have been solved at many levels of approximation on modern computers. Schematically, they are solved in the following manner. Assume there are a finite number of levels, and that the highest excited level is level m. By fiat, level m can only be populated by direct electron recapture because no higher levels exist from which cascading can occur. Therefore, the equilibrium condition for level m requires that

$$N_e N_{HII} \langle \sigma_m^{rec} v \rangle = N_m \sum_{n'=2}^{m-1} A_{mn'}, \quad (3.37)$$

or, solving for N_m,

$$N_m = N_e N_{HII} \frac{\langle \sigma_m^{rec} v \rangle}{A_m} = N_e N_{HII} f_m(T_e), \quad (3.38)$$

$$\text{where } A_m \equiv \sum_{n'=2}^{m-1} A_{mn'}, \text{ and } f_m(T_e) \equiv \frac{\langle \sigma_m^{rec} v \rangle}{A_m} \propto T_e^{-1/2}. \quad (3.39)$$

Thus, N_m is known in terms of the gas density N, since $N_e N_{HII} = N^2$, assuming the gas to be pure H and completely ionized, and that $T_e = 10^4$ °K. Note that $N_m \propto N^2 T_e^{-1/2}$; i.e., it is directly related to the recombination rate. With N_m

now known, one can proceed to the next lower level, and solve for N_{m-1},

$$N_e N_{HII} <\sigma^{rec}_{m-1} v> + N_m A_{m,m-1} = N_{m-1} A_{m-1}, \tag{3.40}$$

or,

$$N_{m-1} = \frac{1}{A_{m-1}} [N_e N_{HII} <\sigma^{rec}_{m-1} v> + N_e N_{HII} f_m(T_e) A_{m,m-1}], \tag{3.41}$$

$$= N_e N_{HII} f_{m-1}(T_e),$$

$$\text{where } f_{m-1}(T_e) = \frac{<\sigma^{rec}_{m-1} v> + <\sigma^{rec}_m v> \frac{A_{m,m-1}}{A_m}}{A_{m-1}} \propto T_e^{-1/2}. \tag{3.42}$$

This procedure can be continued to lower levels. The solutions obtained are all similar, and can generally be written in the form $N_n = N_e N_{HII} f_n(T_e)$, where $f_n(T_e) \propto T_e^{-1/2}$.

The solutions of the capture-cascade equations can now be used to determine the relative intensities of the H lines. We will consider the Balmer lines only, since they are in the visible region of the spectrum and are observed to be relatively bright in most nebulae. The relative fluxes of the Balmer lines, normalized to $F_{H\beta} = 100$, are called the "Balmer decrement". Because the level populations N_n all have the same dependence upon density and temperature, the emission coefficients can be removed from the flux integral (equation 3.32) in the expression for the Balmer decrement, and one has the result that

$$\frac{F_{k,2}}{F_{H\beta}} = \frac{\frac{4\pi}{R^2} \int_o^R j_{k,2}(r)\, r^2 dr}{\frac{4\pi}{R^2} \int_o^R j_{H\beta}(r) r^2 dr} = \frac{j_{k,2}}{j_{H\beta}} = \frac{N_k A_{k,2} \nu_{k,2}}{N_4 A_{H\beta} \nu_{H\beta}} \tag{3.43}$$

$$= \frac{N_e N_{HII} f_k(T_e)\, A_{k,2}\, \nu_{k,2}}{N_e N_{HII} f_4(T_e)\, A_{H\beta}\, \nu_{H\beta}} \propto \frac{f_k(T_e)}{f_4(T_e)} \propto \text{const} X\, \frac{T_e^{-1/2}}{T_e^{-1/2}} = \text{const}(k),$$

where k is any level $k > 2$. That is, the relative intensities of the Balmer lines are essentially independent of T_e and N, and all other physical parameters of the gas. The normal Balmer decrement for a nebula which is optically thick in the Lyman lines has been tabulated by a number of people (see Brocklehurst 1971), and is given in Table 3.5.

The fact that the Balmer line intensities are insensitive to physical conditions in the gas means that the decrement cannot be used to extract this information from nebulae. On the other hand, the constancy of the decrement can be used as a check on recombination theory and intrinsic or interstellar reddening. The observed Balmer decrement in most nebulae agrees fairly well with the theoretical predictions. In some objects there is some disagreement, but in all instances it appears that this can be ascribed to either observational uncertainty or reddening. In fact, correcting the observed Balmer lines to their theoretical values is the standard method of determining the reddening of individual nebulae.

B. Absolute Flux of Hβ . Consider a spherical nebula of radius R, at a distance $d \gg R$, which has an observed angular diameter $2\theta_o = 2R/d$. The emission coefficient for Hβ radiation, $j_{H\beta}$, represents the energy rate of production of Hβ photons per steradian per unit volume of gas, therefore it is simply related to the Hβ luminosity of the nebula,

$$L_{H\beta} = 4\pi j_{H\beta}\, 4/3\pi R^3 \qquad \text{ergs/sec}, \qquad (3.44)$$

assuming a homogeneous gas which is optically thin to Hβ , i.e., all created Hβ photons escape directly. If we use the usual relationship between flux and luminosity, $L_{H\beta} = 4\pi d^2 F_{H\beta}$, where $F_{H\beta}$ is the observed Hβ flux, then we can write that

TABLE 3.5

The Hydrogen Balmer Decrement ($F_{H\beta}$=100)

Line \ T_e / N_e	10,000°		20,000°	
	10^2 cm^{-3}	10^6 cm^{-3}	10^2 cm^{-3}	10^6 cm^{-3}
Hα	286	281	274	272
Hβ	100	100	100	100
Hγ	47	47	48	48
Hδ	26	26	26	27
Hε	16	16	16	16
Hζ	11	11	11	11
H10	5.3	5.9	5.4	5.7
H15	1.6	2.1	1.6	1.8

$$F_{H\beta} = 4/3\pi j_{H\beta} \frac{R^3}{d^2} = 4/3\pi N_e{}^2 f_4(T_e)\ A_{H\beta} \frac{h\nu_{H\beta}}{4\pi}\ \theta_o^3\ d, \quad (3.45)$$

where we have assumed a pure hydrogen gas ($N_e = N_{HII}$). The absolute Hβ flux, $F_{H\beta}$, and the angular size of a nebula, θ_o, are always directly observable quantities. The function $f_4(T_e)$ is a known, insensitive function of T_e, which can be assumed to be 10^4 °K, and therefore equation (3.45) may be taken to be a relation between the distance of a nebula and its mean square electron density. If N_e is known from, say, the [OII] λ3727 line (see following discussion), then the Hβ flux can be used to get a rough distance to the nebula. Alternatively, if the distance to a nebula is known, say, from the spectroscopic parallax of the ionizing star, then it is possible to find an average N_e for the gas. Densities obtained in this manner are rms values averaged over the entire volume of the nebula. Such values may differ from the true local value of N_e at any given point in the gas because of density fluctuations.

C. Forbidden Lines of the Heavy Elements. Like the Balmer lines, the forbidden lines are optically thin in gaseous nebulae, so the line fluxes may be obtained directly from the emission coefficients (equation 3.32). The emission coefficient of a line is related to the upper level population, as defined in equation (3.33), and therefore the determination of forbidden line intensities requires knowledge of the excited state population. As in the case of the recombination lines, the level populations are obtained from statistical equilibrium. The dominant processes affecting population of the low-lying levels of the heavy elements are collisional excitation, and collisional and radiative de-excitation. Radiative excitation may be ignored because of the dilute radiation field, and recombination may be ignored because of its much smaller cross section compared with collisional excitation. The general situation is represented by the 2-level atom: ground and upper state, denoted levels 1 and 2, respectively. The steady-state equation for the population of the upper level equates the rate of 1 → 2 collisional excitations/(cm^3 sec) to the rate of 2 → 1 radiative and collisional de-excitations/(cm^3 sec), or

$$N_eN_1 <\sigma_{12}\, v> = N_2(A_{21} + N_e <\sigma_{21}\, v>). \qquad (3.46)$$

The solution of this equation for N_2 and the emission coefficient j_{21} is

$$N_2 = N_eN_1 \frac{<\sigma_{12}v>}{A_{21}+N_e<\sigma_{21}v>}, \qquad (3.47)$$

$$j_{21} = N_2A_{21} \frac{h\nu_{21}}{4\pi} = N_eN_1<\sigma_{12}v> \frac{A_{21}}{A_{21}+N_e<\sigma_{21}v>} \frac{h\nu_{21}}{4\pi}. \qquad (3.48)$$

That is, the rate at which downward radiative transitions occur is just the rate of upward collisional excitations times the fraction of those downward transitions that are radiative. Since most ions are in the ground state, N_1 may simply be taken to be the ion (or atom) density.

Both $\sigma_{12}(v)$ and $\sigma_{21}(v)$ have the same dependence upon velocity, $\sigma_{12}(v) \propto \sigma_{21}(v) \propto 1/v^2$. We have already shown how reaction rates depend upon T_e when $\sigma \propto v^{-2}$. Since collisional excitation has a threshold, $1/2\, m_e v^2 \geq \chi$, $<\sigma_{12}\, v> \propto T_e^{-1/2}\, e^{-\chi/kT_e}$. Collisional de-excitation has no threshold, therefore $<\sigma_{21}\, v> \propto T_e^{-1/2}$. In fact, since collisional excitation is the inverse process to de-excitation, the principle of detailed balancing can be used to show that the excitation and de-excitation coefficients must be related by

$$<\sigma_{12}\, v> = <\sigma_{21}\, v> \frac{g_2}{g_1} e^{-\chi/kT_e}, \qquad (3.49)$$

where g_1 and g_2 are the statistical weights of the two levels.

Consider the expressions for N_2 and j_{21} in the limits that $A_{21} \gtrless N_e <\sigma_{21}\, v>$. The density at which collisional de-excitation and radiative de-excitation occur equally is called the "critical density", N_e^c, for that transition,

$$N_e^c \equiv \frac{A_{21}}{<\sigma_{21}v>}. \qquad (3.50)$$

For forbidden lines, typical transition probabilities are $A_{21} \sim 1\ sec^{-1}$. At temperatures of 10^4°K, for most of the lower levels of ions of the heavy elements, $<\sigma_{21}> \sim 10^{-16} cm^2$, and $<v> \sim 10^8$ cm/sec for the electrons. Therefore, $N_e^c \sim 10^8 cm^{-3}$ for most forbidden transitions. Densities in nebulae are lower than this, so collisional de-excitation is only important for a few transitions having unusually small transition probabilities. The [OII] λ3727 doublet, which will be considered shortly, is a good example of a line with a low critical density.

(1) "Low density limit". The limit in which a level is predominantly radiatively de-populated, i.e., $A_{21} >> N_e <\sigma_{21} v>$, is referred to as the low density limit. When this condition is satisfied,

$$j_{21} = N_e N_1 <\sigma_{12} v> \frac{h\nu_{21}}{4\pi} , \quad (3.51)$$

and

$$\frac{N_2}{N_1} = N_e \frac{<\sigma_{12} v>}{A_{21}} . \quad (3.52)$$

In the limit that there is essentially no collisional de-excitation, the line emission does <u>not</u> depend upon A_{21}. Every $1 \to 2$ collisional excitation leads to a $2 \to 1$ radiative de-excitation, independent of the lifetime of the level. Furthermore, in the low density limit, the emission coefficient depends upon the square of the density, $j_{21} \propto N^2$.

(2) "High density limit". A level which is predominantly collisionally de-excited, $N_e <\sigma_{21} v> >> A_{21}$, is in the high density limit. The emission coefficient and level population are then

$$j_{21} = N_1 A_{21} \frac{<\sigma_{12} v>}{<\sigma_{21} v>} \frac{h\nu_{21}}{4\pi} , \quad (3.53)$$

$$\frac{N_2}{N_1} = \frac{<\sigma_{12} v>}{<\sigma_{21} v>} = \frac{g_2}{g_1} e^{-\chi/kT_e} . \quad (3.54)$$

That is, when most collisional excitations lead to collisional, rather than radiative, de-excitations, the emission does depend upon A_{21}. Also, there is a diminished dependence of j_{21} on the density, $j_{21} \propto N$, rather than N^2. And, when collisions dominate, the level populations approach Boltzmann values. As the rate of collisional excitation is increased by increasing N, most of these atoms are collisionally returned to the ground state, and do not add to the radiation field.

As an illustrative example, consider the relative intensities of two lines, one permitted (A_{21} large) and the other forbidden (A_{21} small), from the same ion as a function of increasing gas density N in an ionized gas which has a kinetic temperature T. Initially, at low density, both lines are in the low density limit, and the intensities of both increase as N^2, independent of the individual Einstein coefficients. Assuming the collision cross sections of the levels and the line frequencies to be similar, the two lines will have comparable intensities (cf. equation 3.51), as shown in Figure 3.4. As the density increases, the critical density of the forbidden line (N_F) is eventually reached, and its upper level becomes predominantly collisionally de-excited. At this point, the permitted line is still in the low density limit, because its critical density is much higher than that for the forbidden line. For higher densities $N > N_F$, the intensity of the forbidden line increases only as N, whereas the permitted line intensity still increases as N^2. The ratio of the permitted/forbidden line intensity $\propto$ N, and rapidly increases. Eventually, at higher densities the permitted line also reaches its high density limit ($N = N_P$), and its intensity $I_P \propto N$. The intensity ratio of the two lines then remains constant $I_P/I_F \propto A$ (permitted) /A(forbidden) ~ 10^8, independent of density. This situation does not remain true for arbitrarily higher densities, however. The intensity of a line is proportional to the emission coefficient only for an optically thin gas (see Chap. 1). When the gas becomes optically thick at a wavelength, the intensity is then equal to the source function, which for thermal radiation is the Planck function, $B_\nu(T)$. The optical depth of a gas at the wavelength of a line is directly proportional to the density and size of the gas, and the Einstein coefficient A_{21} of the line. Consequently, as the density of the gas increases in our thought experiment, assuming the size to remain constant, the point

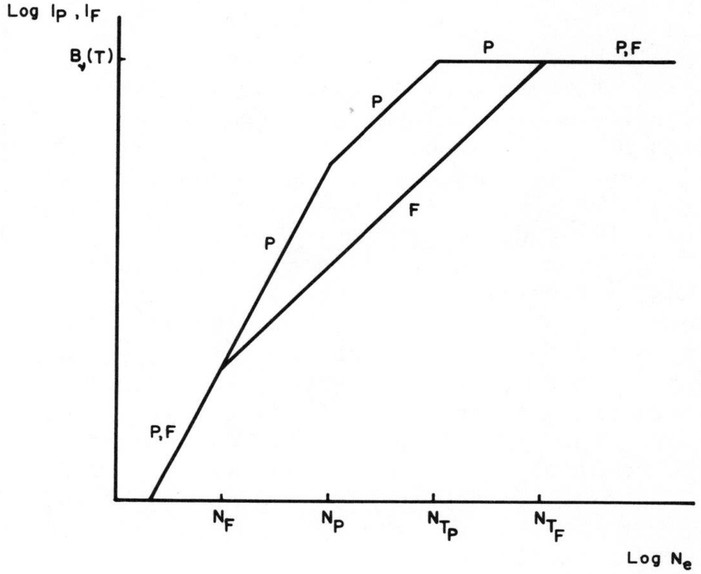

Fig. 3.4 A comparison of the absolute intensities of two emission lines in an ionized gas as a function of the electron density N_e. The lines are assumed to be identical in all respects except for widely differing spontaneous transition probabilities A_{21}. For densities $N_e < N_F$, where N_F is the critical density of the forbidden line (typically $N_F \sim 10^8$ cm^{-3}), the two lines have equal intensities. For $N_F < N_e < N_P$, where N_P is the critical density of the permitted line (typically $N_P \sim 10^{16}$ cm^{-3}), the permitted line intensity $I_P \propto N_e^2$, whereas the forbidden line intensity $I_F \propto N_e$. Consequently, the permitted line is much brighter in this density regime. It remains brighter even after densities are reached where it is collisionally de-excited, $N > N_P$. Eventually, a density is reached where the permitted line becomes optically thick, $N_e = N_{TP}$, and its intensity becomes equal to the Planck function. The intensity of the forbidden line continues to increase until it, too, becomes optically thick, at $N_e = N_{TF}$, and for all $N_e > N_{TF}$, the intensities are equal: $I_F = I_P = B_\nu(T)$.

will eventually be reached where the gas becomes optically thick, and the line intensity will no longer increase with the density, but will approach a limiting value which is the Planck function for that frequency and gas temperature $I = B_\nu(T)$. Conceivably, a gas could become optically thick to a line at a density for which the line is still in the low density limit, however, this is usually not the case. As the density increases above the critical density of the permitted line N_P, (for typical permitted lines, $N_P \sim 10^{16} cm^{-3}$) the intensity ratio of the permitted/forbidden line remains constant, until a density is reached where the permitted line becomes optically thick. At this point the permitted line intensity becomes independent of density, whereas the intensity of the forbidden line, which is not yet optically thick because of its smaller Einstein coefficient, still increases linearly with the density. Ultimately, a density is reached where the forbidden line also becomes optically thick, and its intensity also reaches the Planck function at the kinetic temperature of the gas. For densities greater than this, the intensities of the permitted and forbidden lines are again equal.

According to the above analysis, forbidden and permitted lines have widely different intensities only when they are both optically thin and the gas density is such that the permitted line is in the low density limit while the forbidden line is in the high density limit ($10^8 \lesssim N \lesssim 10^{16} cm^{-3}$, typically). Confined plasmas in terrestrial laboratories fulfill these conditions, hence forbidden lines are very weak compared with permitted lines. However, the low densities generally encountered in gaseous nebulae cause forbidden lines and permitted lines both to be in the low density limit, and it is for this reason they have comparable intensities in nebulae.

Unlike the Balmer lines, the relative intensities of the forbidden lines may be used to determine physical conditions in the emitting gas. Specifically, it is possible to use relative intensities of certain lines to deduce the electron temperature and density of the nebulae.

(1) Determination of T_e. Because the upper levels of all forbidden lines are populated by collisional excitation from the ground state by free electrons, the intensities of the forbidden lines are dependent upon the electron temperature. Higher levels are populated less rapidly than lower levels because of the higher threshold energy required of the exciting electrons, and a comparison of two

collisionally excited lines from the same ion enables a straightforward determination of the electron temperature to be made, independent of the abundances of the elements or ionization of the gas.

The atomic structure of ions with a $1s^2\ 2s^2\ 2p^2$ and $1s^2\ 2s^2\ 2p^4$ electron configuration is such that several excited levels lie within a few volts of the ground state. The OIII ion is the most abundant of the ions in nebulae having this structure, and the relevant lines from transitions among these levels happen to fall in the visible portion of the spectrum, so the [OIII] forbidden lines are usually used to find T_e in nebulae. An energy level diagram of the lower levels of OIII is given in Figure 3.5, with the appropriate transitions labelled. The ^3P ground state is split into fine structure levels. However, for purposes of this discussion we shall treat the ground state as one level, and consider the λ5007 and λ4959 lines to be one line: λ5007. The λ4363 and λλ5007, 4959 lines all have Einstein coefficients $A \gtrsim 1\ \mathrm{sec}^{-1}$, therefore collisional de-excitation may be ignored for these levels at nebular densities. Every $1 \rightarrow 2$ or $1 \rightarrow 3$ collisional excitation upward leads to a downward $3 \rightarrow 2$, $3 \rightarrow 1$, or $2 \rightarrow 1$ radiative de-excitation. The emission coefficients for the λ4363 and λ5007 lines may be written

$$j(\lambda 4363) = N_e N_1 \langle\sigma_{13} v\rangle \frac{h\nu_{32}}{4\pi} \frac{A_{32}}{A_{31} + A_{32}} , \qquad (3.55)$$

and

$$j(\lambda 5007) = N_e N_1 [\langle\sigma_{12} v\rangle + \langle\sigma_{13} v\rangle \frac{A_{32}}{A_{31} + A_{21}}] \frac{h\nu_{21}}{4\pi} . \qquad (3.56)$$

The branching ratio $A_{32}/(A_{31} + A_{32}) \sim 1$, since $A_{32} >> A_{31}$, i.e., most $1 \rightarrow 3$ excitations produce a λ4363 photon. After the OIII has decayed to level 2 by producing λ4363, it then returns to the ground state by emission of λ5007. Therefore, λ5007 can be created by either $1 \rightarrow 2$ or $1 \rightarrow 3$ collisional excitations, and the two terms in equation (3.56)

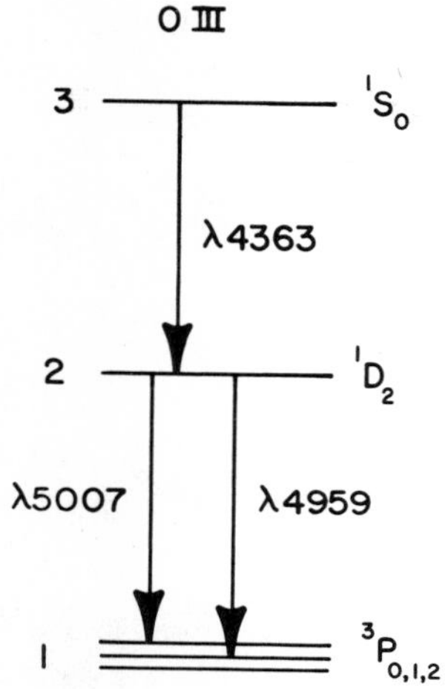

Fig. 3.5 Energy level diagram for the lowest levels of OIII

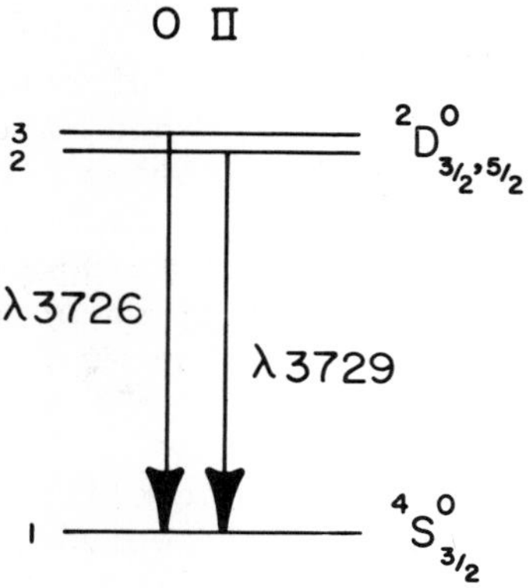

Fig. 3.6 Energy level diagram for the lowest levels of OII

represent each of these possibilities. For optically thin lines (in an idealized homogeneous gas), the intensity ratio of λ5007 to λ4363 may be written

$$\frac{I(\lambda 5007)}{I(\lambda 4363)} = \frac{j(\lambda 5007)}{j(\lambda 4363)} = \frac{N_e N_1 [\langle\sigma_{12} v\rangle + \langle\sigma_{13} v\rangle] \frac{h\nu_{21}}{4\pi}}{N_e N_1 \langle\sigma_{13} v\rangle \frac{h\nu_{32}}{4\pi}}, \quad (3.57)$$

$$= \frac{\lambda_{32}}{\lambda_{21}} [1 + \frac{\langle\sigma_{12} v\rangle}{\langle\sigma_{13} v\rangle}], \quad (3.58)$$

where $\langle\sigma_{12}\ v\rangle \propto T_e^{-1/2} \exp[-\chi_{12}/kT_e]$ and $\langle\sigma_{13}\ v\rangle \propto T_e^{-1/2} \exp[-\chi_{13}/kT_e]$. Since $\langle\sigma_{12}\rangle \sim 10 \langle\sigma_{13}\rangle$ and $\chi_{12} < \chi_{23}$, the direct collisional excitation of level 2 is predominant in the creation of λ5007, and therefore the left-hand term may be neglected, with the final result being

$$\frac{I(\lambda 5007)}{I(\lambda 4363)} \simeq 8e^{33,000/T_e}. \quad (3.59)$$

Thus, observations of the [OIII] lines give T_e directly. Physically, the reason for this is that λ4363 and λ5007 originate from different upper levels, and therefore have different excitation thresholds, leading to a different T_e-dependence. This method is very good for getting temperatures of the emitting gas, although it is sometimes difficult to determine I(λ4363) accurately. Because of its high excitation potential, λ4363 tends to be weak at nebular temperatures.

(2) Determination of N_e. The emission coefficient of a line in the high density limit has a different dependence upon the electron density than that of a line in the low density limit. Hence, a comparison of two lines, one

of which is predominantly collisionally de-excited and the other which is primarily radiatively de-excited, will be dependent upon N_e. If the two lines come from different levels or different ions, their relative intensities will also depend upon the temperature and ionization of the gas. However, if they originate from the same upper level of an ion, their relative strengths will depend only on the density. The low-lying excited levels of OII, which has a $1s^2 2s^2 2p^3$ electronic configuration (see Figure 3.6), are split by fine structure, and they have differing Einstein coefficients. The lines from these levels fall in the visible and are fairly bright in most nebulae, therefore they are very useful for determining the density in nebulae.

The $^4S^o_{3/2}$ level is the ground state of OII. The fine-structure splitting of the excited levels is very small, so levels 2 and 3 have essentially the same energy. The transitions between the first excited state and the ground state, λ3726 and λ3729, are forbidden, and are usually referred to as the [OII] λ3727 doublet. The transition probabilities of the lines are $A(\lambda3729) = 4 \times 10^{-5}$ sec^{-1}, and $A(\lambda3726) = 2 \times 10^{-4}$ sec^{-1}. These values are relatively small, consequently both lines have critical densities which are comparable to the normal densities of planetary nebulae. It has been shown from atomic physics calculations that $\sigma_{21}(v) = \sigma_{31}(v)$, where $\langle\sigma_{21}\rangle = \langle\sigma_{31}\rangle \simeq 10^{-16}$cm^2 for a temperature of 10^4°K. Substituting the above transition probabilities into equation (3.50) leads to the following values for the critical densities of the two lines: $N_e^c(\lambda3729) = 4 \times 10^3$cm^{-3}, and $N_e^c(\lambda3726) = 2 \times 10^4$ cm^{-3}. Consider the intensity ratios of these two lines in the high and low density limits of both lines:

(a) $N_e \lesssim 4 \times 10^3$cm^{-3} (both lines in the low density limit)

$$\frac{I(\lambda3729)}{I(\lambda3726)} = \frac{j_{21}}{j_{31}} = \frac{N_e N_1 \langle\sigma_{12}v\rangle \frac{h\nu_{21}}{4\pi}}{N_e N_1 \langle\sigma_{13}v\rangle \frac{h\nu_{31}}{4\pi}} = \frac{\langle\sigma_{12}v\rangle}{\langle\sigma_{13}v\rangle}, \qquad (3.60)$$

assuming that collisional interactions between the fine-structure levels are unimportant. It is generally true that collision cross sections to fine structure levels differ only by the statistical weights of the levels, therefore $\sigma_{12}(v)/\sigma_{13}(v) = g_2/g_3$, where g_2 and g_3 are the statistical weights of levels 2 and 3. Since $g = 2J+1$, where J is the quantum number denoting the total angular momentum of the level ($J_2 = 5/2$, $J_3 = 3/2$), $g_2/g_3 = 1.5$. So, $I(\lambda3729)/I(\lambda3726) = 1.5$ when both lines are in the low density limit - the intensity ratio is a constant independent of <u>any</u> physical conditions in the gas.

(b) $N_e \gtrsim 2 \times 10^4 cm^{-3}$ (both lines in the high density limit)

$$\frac{I(\lambda3729)}{I(\lambda3726)} = \frac{j_{21}}{j_{31}} = \frac{N_1 A_{21} \frac{<\sigma_{12}v>}{<\sigma_{21}v>} \frac{h\nu_{21}}{4\pi}}{N_1 A_{31} \frac{<\sigma_{13}v>}{<\sigma_{31}v>} \frac{h\nu_{31}}{4\pi}} = \frac{A_{21}}{A_{31}} \frac{g_2}{g_3} = 0.3. \quad (3.61)$$

That is, when both lines are in the high density limit, their relative intensities are again constant, independent of density or temperature.

(c) $4 \times 10^3 \lesssim N_e \lesssim 2 \times 10^4 cm^{-3}$ (λ3726 in the low density limit, λ3729 in the high density limit).

When one line is in the high density limit while the other is in the low density limit, their intensities have different dependences upon N_e, and the relative strengths of the lines is therefore density-dependent. In this density regime,

$$\frac{I(\lambda3729)}{I(\lambda3726)} = \frac{N_1 A_{21} \frac{<\sigma_{12}v>}{<\sigma_{21}v>} \frac{h\nu_{21}}{4\pi}}{N_e N_1 <\sigma_{13}v> \frac{h\nu_{31}}{4\pi}} = \frac{g_2}{g_3} \frac{A_{21}}{N_e <\sigma_{21}v>} ,$$

$$\simeq \text{const} \times \frac{T_e^{1/2}}{N_e} . \quad (3.62)$$

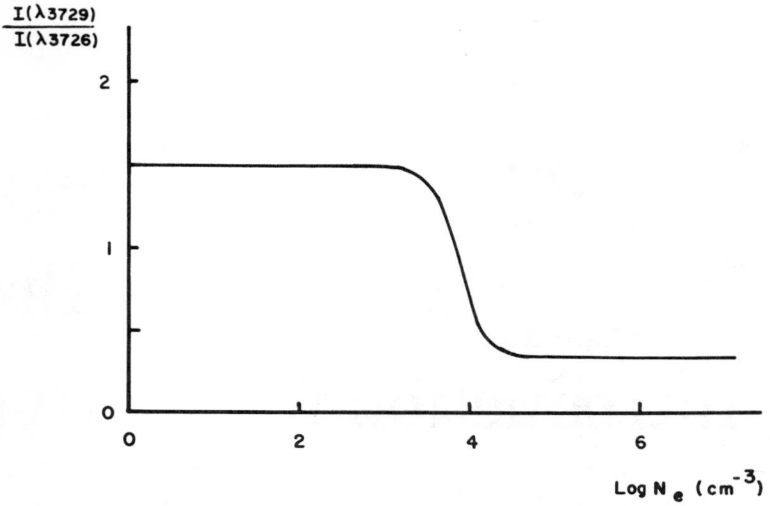

Fig. 3.7 The intensity ratio of the lines of the [OII] λ3727 doublet as a function of electron density

When the density is such that one level is predominantly collisionally de-excited while the other is radiatively de-excited, the intensity ratio depends directly upon N_e, with a slight temperature dependence. A schematic plot of the intensity ratio of the λ3727 doublet is given in Figure 3.7, after Osterbrock (1955). Only when the density is in the range of $N_e \sim 10^3$-$10^4 cm^{-3}$ can N_e be obtained from the [OII] lines. Outside of this range, only limits can be found to the density. Fortunately, the densities of most planetaries fall in the range where λ3727 may be used to obtain N_e. However, it does require spectrometry of sufficient resolution to resolve the two lines.

Limited space does not allow a discussion of several topics which are of equal importance to any of those covered above. However, a comprehensive text has recently been published which may be consulted by anyone interested in a more thorough discussion of the physical processes in gaseous nebulae (Osterbrock 1974).

4. A Brief Introduction to Relativity

4.1 Introduction

This Chapter is intended to serve as a brief, but rigorous and uncondensed, introduction to special relativity for the advanced astronomy student. We find it advantageous to cover the companion topic, tensor analysis, in such a way that the student will be able to make the transition to general relativity and cosmology with as little pain as possible. Indeed, connections with general relativity are mentioned here quite often.

The reader is assumed to be familiar with coordinate transformations in general and, for the sections on tensor analysis, to have a reasonable grasp of advanced calculus. The section on Maxwell's equations and particle dynamics (Section 4.5) cannot be understood without some understanding of vector analysis. It is also assumed that the reader is familiar with the simple "standard Lorentz transformation," although he may be unsure of how to use it.

Many of the standard topics are covered here, such as the Lorentz force equation, Maxwell's equations, aberration, and the Doppler effect. Also included are a few miscellaneous applications of astrophysical interest. We do not treat collision theory or relativistic hydrodynamics.

At this point, we might make a few remarks about notation and about our basic point of view. The basic geometric entity in special relativity is the event, which means a single point (x, y, z, t) in the four-dimensional continuum of space and time. We denote it by the symbol P. If there are two events in question, P_1 and P_2, then x_1 means the x coordinate of the first event and x_2 the x coordinate of the second event, etc. The Lorentz transformation (LT) is then a transformation of space-time coordinates between coordinate systems which are moving relative to each other. The LT relates the coordinates of some fixed, but arbitrary, event as measured by one coordinate system (the 0 system), to the coordinates <u>of that same event</u> as measured by the second coordinate system (the 0' system). If we are interested in an event, P_1, say the explosion of a firecracker, then (x_1, y_1, z_1, t_1) will denote the space-time coordinates of the explosion as seen from the standpoint of the 0 system, and (x_1', y_1', z_1', t_1') will denote the coordinates of the explosion as seen from the 0' system. The Lorentz Transformation for the coordinates of this event is $x_1' = (x_1 - Vt_1)/(1-\beta^2)^{1/2}$, $y_1' = y_1$, $z_1' = z_1$, $t_1' = (t_1-\beta x_1/c)/(1-\beta^2)^{1/2}$, where V is the relative velocity between the 0 and 0' systems, c is the velocity of light, and β means V/c. Thus the LT finds the 0' coordinates of our fixed event as functions of the 0 coordinates of that event.

In general, the author has found that it is better for the student to learn the basic principles of relativity thoroughly and then go on to some advanced applications, rather than to try to cover many applications with only a superficial understanding. It is hoped that the present treatment is consistent with this approach.

4.2. Hypotheses Underlying the Lorentz Transformation

The Lorentz Transformation (LT) is discussed and derived in many texts, primarily in those concerned with relativity per se, but also in texts on electrodynamics and on classical mechanics. Therefore, we do not present a derivation here, but rather give a sketch of the hypotheses which underly it.

We assume that we deal only with transformations between "standard inertial coordinate systems" (SICS's) which are uniformly moving with respect to each other, in which spatial (three-dimensional) geometry is Euclidean, and in which freely moving bodies appear to be moving in uniform rectilinear motion. Thus we have the following definition: An SICS is a rectangular Cartesian coordinate system in which a body moving with no external applied forces appears to move in a straight line at a uniform rate. This makes sense only if some uniformly running system of clocks is used to measure time throughout the SICS. It is important to realize that the Lorentz transformation (LT) can be "derived" only if it is required that the coordinate systems in the two SICS's be Cartesian, and that the units of length and time be chosen in the same way in both. Therefore, we must introduce the important hypothesis that any two SICS's have complete geometrical and physical reciprocity with respect to each other, so that any experiment performed in the first SICS will have exactly the same result if it is performed in an identical way in the second SICS. This reciprocity hypothesis is meant only in the most obvious, common-sense sort of way, and in particular, has the consequence that if the first SICS is observed by the second to move with a certain speed v, then the second will be observed by the first to move at the same speed, v, although obviously the velocities will be opposite to each other.

The thoughtful reader may realize that there is quite a bit of physics hidden in this hypothesis. For example, to measure velocities, observers in both frames of reference will need similarly constructed time and length standards. Length standards might be set up by, say, choosing a unit of length to be that of a salt crystal so-and-so many atoms long on an edge. Each observer could then fabricate his own length standards by himself in his own SICS. The reciprocity hypothesis would then apply to measurements made with these length standards. Having set up length standards, it would then be easy to set up time standards by agreeing to use clocks constructed in a certain specified way out of standard materials.

Also very important is the hypothesis of spatial and temporal homogeneity and isotropy: Space-time is assumed to be homogeneous and isotropic. Thus all points (events) in space-time are equivalent, as are all directions in space.

It is convenient at this point to define the proper length of an object as the length as measured in an SICS with respect to which the object is at rest. Then, we can interpret the reciprocity hypothesis as implying that similarly constructed measuring rods have the same proper

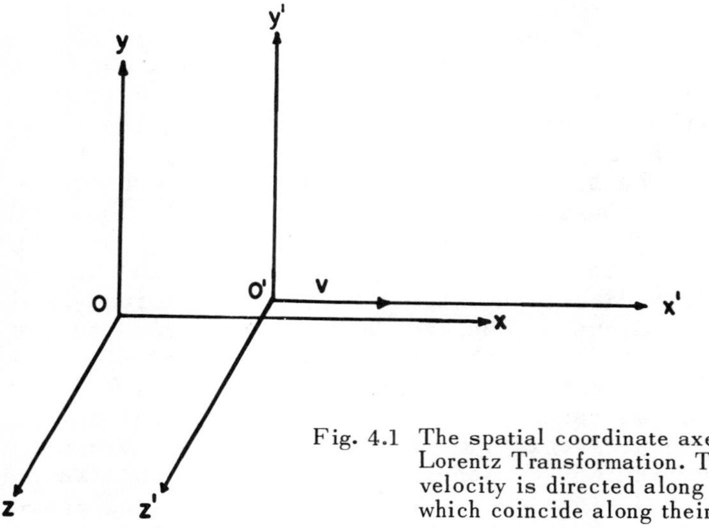

Fig. 4.1 The spatial coordinate axes for the Standard Lorentz Transformation. The transformation velocity is directed along the x and x' axes, which coincide along their lengths.

lengths as measured in different SIC's. It also implies that length and time standards may be accelerated, perhaps very gently, from one SICS to another, and compared with standards constructed there according to the same prescription. They will then be observed to be similarly constructed and therefore to give the same results.

The above two hypotheses (reciprocity, and homogeneity and isotropy) are compatible with both the Galilean transformation and the LT, and one can use them to show that it is possible to orient and translate the axes in any two SICS's such that the y and z axes are mutually parallel, and such that the x-axes coincide along their entire lengths. See Figure 4.1. To get the LT, we introduce <u>Einstein's Principle of Relativity: The reciprocity of SICS's extends to all the laws of mechanics and electromagnetism, in such a way that the velocity of light</u>, $c = 2.998...\times 10^{10}$ cm/sec, <u>is the same as measured in all SICS's, independent of the velocity of the source</u>.

One can then show that the LT follows; that is,

$$x' = \frac{x - Vt}{(1-\beta^2)^{1/2}}, \quad y' = y, \quad z' = z, \quad t' = \frac{t - \beta x/c}{(1-\beta^2)^{1/2}}, \tag{4.1}$$

where $\beta \equiv V/c$ is the transformation velocity measured in units of the velocity of light.

This transformation, with the spatial axes all aligned and the transformation velocity along the x axis, is called the Standard Lorentz Transformation (SLT). This particular configuration of 0 and 0' axes is called the Standard Configuration.

It should be stated that the most general homogeneous LT can be compounded out of two spatial rotations and one SLT, and possibly space or time inversions. First the 0 system is rotated so that the x axis lies along the transformation velocity direction, then the SLT is performed, and 0' is then rotated out of the standard configuration to its final desired orientation. The possible space and time inversions may be interposed at any stage.

The Jacobian determinant of the SLT is equal to +1, as the reader should verify. If no inversions are performed, the Jacobian of the general LT is +1, and the transformation is called a proper LT. If an odd number of space inversions occur, and/or if a time inversion occurs, the transformation is termed improper. In the latter case, it is impossible to construct the total LT out of proper spatial rotations together with the SLT.

At the risk of seeming repetitious, we emphasize again that the LT is a relation between coordinates of a fixed, single event in space time. The reader might wish to review the remarks about this in the Introduction.

As an example, let us apply the LT to explosive outbursts in a quasi-stellar object (QSO), moving with a speed close to the velocity of light, say $V = 3c/4$, so that $\beta = 3/4$ and $(1-\beta^2)^{-1/2} \equiv \Gamma = 4/\sqrt{7} \simeq 1.51$. We suppose that the QSO is in our immediate neighborhood, so that we can ignore cosmological effects, which have to be treated with a gravitation theory, such as the general theory of relativity. We suppose that the QSO is moving directly away from us and that we have two outbursts that took place one year apart as measured in the rest frame of the QSO, which we call the 0' system. We thus deal with two events, P_1 (the first outburst) and P_2 (the second outburst). Let the origin of 0' be at the QSO, and let the QSO be at a distance $x_1 = L$ from us at the time of the first outburst. Then $x_1' = 0$, and the first of equations (4.1), with the abbreviation $(1 - \beta^2)^{-1/2} \equiv \Gamma$, implies $\Gamma(L - Vt_1) = 0$, or $L = Vt_1$. Thus, the distance traveled by the QSO since the origins were coincident is equal to the velocity multiplied by the elapsed time, in the usual way.

We now ask about the time at which the outburst is actually detected by us at our origin, 0. This time will clearly be greater than $t_1 = L/V$, since the light (or radio waves) signaling the outburst requires an additional time L/c to travel back to us. Thus the observed time is $t_1^{obs} = t_1 + x_1/c = L(V^{-1} + c^{-1})$. This is not the coordinate time for the event, which has been already measured by an 0 clock at the event itself.

What is t_1'? We can use the last of equations (4.1) to get $t_1' = \Gamma L(V^{-1} - \beta c^{-1}) = L/(\Gamma V) = t_1/\Gamma$. This result is harder to understand, but we will tie it in later (Section 4.3) with the time dilatation phenomenon.

For the second event P_2, we know $t_2' - t_1' = \Delta t' =$ 1 year, from the statement of the problem. Also, $x_2' = 0$, and therefore $x_2 = Vt_2$. Substitution into the last of equations (4.1) gives $t_2' = \Delta t' + L/(\Gamma V) = t_2/\Gamma$, or $t_2 = \Gamma\Delta t' + L/V$. Since $L/V = t_1$, we have $t_2 - t_1 = \Delta t = \Gamma\Delta t' \simeq 1.51$ years, remembering $\Gamma \simeq 1.51$. This relates the coordinate time differences between the two events. We must now find x_2, by noting $x_2 = Vt_2 = V\Gamma\Delta t' + L$, from the above equations, and this enables us to compute the corresponding observed time at 0, by writing $t_2^{obs} = t_2 + x_2/c = \Gamma\Delta t' + L/V + \beta\Gamma\Delta t' + L/c$. The difference between observed times (still at the origin of 0) is then $\Delta t^{obs} = t_2^{obs} - t_1^{obs} = \Gamma(1+\beta)\Delta t' = [(1+\beta)/(1-\beta)]^{1/2}\Delta t' = 2.65$ years.

This time interval is not the coordinate time interval between P_1 and P_2, which is 1.51 years in the 0 system, but is related to the Doppler effect, which we discuss in Section (4.7). It is important to realize that in interpreting light signals coming from events, the light travel times must always be allowed for in computing coordinate times. It is the coordinate times that are used in the LT.

3. Kinematic and Algebraic Properties of the Lorentz Transformation

We must now discuss the concept of simultaneity in special relativity. Since by Einstein's principle the velocity of light does not depend on direction in an SICS, light signals can be used to synchronize clocks over large distances. Two clocks at rest in an SICS will be called synchronized if an observer at rest situated equi-distantly between them reads (by means of a telescope, perhaps) the same time on both of them continuously. Two events at different locations in an SICS will be called <u>simultaneous with respect to the SICS</u> if they occur at the same time as read by synchronized clocks at their respective locations. As will be shown below, events which are simultaneous with respect to one SICS are not necessarily simultaneous with respect to another.

With the above discussion in mind, consider the measurement in the 0 system of the length of a rod at rest with respect to the 0' system and laid parallel to the x' axis. We assume the 0 and 0' systems to be in the standard configuration, so that the SLT equations (4.1) apply. This measurement is accomplished in 0, with respect to which the rod is moving, by marking the x coordinates of the ends of the rod <u>simultaneously</u> at some time t_o. Consider these marking events as P_1 and P_2. Then the proper length of the rod, which must be measured in the 0' system, is $L_p = x_2' - x_1'$, and the length measured in 0 is $L = x_2 - x_1$. Then the transformation equation $x' = \Gamma(x - Vt)$ yields

$$L_p = x_2' - x_1' = \Gamma[x_2 - x_1 - V(t_2 - t_1)] = \Gamma L, \qquad (4.2)$$

since $t_2 = t_1 = t_o$. Thus the measured length L is shorter than the proper length, since $\Gamma \geq 1$. This is the Lorentz contraction effect. An observer in 0', however, will claim that the marking events P_2 and P_1 did not occur at the same time. Let $\Delta t' \equiv t_2' - t_1'$. Then equations (4.1) imply

$$\Delta t' = \Gamma[t_2 - t_1 - (x_2 - x_1)\beta/c] = -\Gamma L\beta/c = -L_p\beta/c. \qquad (4.3)$$

This is the <u>desynchronization formula</u>. Events that occur simultaneously in 0, and that are separated by an x distance L, appear out of synchrony in 0' by this amount. Thus the Lorentz contraction formula (4.2) related to events that occur <u>at the same time in 0</u>, but at different spatial locations.

For an interesting discussion of thought experiments in which the Lorentz contraction and desynchronization formulas are derived from first principles, the reader is referred to the book by Panofsky and Phillips, Section 16-2.

It is also instructive to inquire about the time and space differences in 0 between two events which occur <u>at the same place in 0'</u>, but at different times. Call these two events P_1 and P_2. Then $x_1' = x_2'$, $\Delta t' \equiv t_2' - t_1'$, and the SLT equations (4.1) imply $x_2 - x_1 = \Gamma V \Delta t'$, and

$$t_2 - t_1 = \Delta t = \Gamma \Delta t' . \tag{4.4}$$

Again, since $\Gamma \geq 1$, we can say that a clock at rest in 0' seems to run slow compared with synchronized clocks at rest in 0. This is the <u>time dilatation effect</u>.

The Lorentz contraction formula (4.2) and time dilatation formula (4.4) refer to measuring rods and clocks which are at rest in 0'.

We have already encountered the time dilatation effect in the example of the QSO outbursts analyzed in Section (4.2). In fact, equation (4.4) has already been derived there. The two outbursts were assumed to happen in the same place in 0'; i.e., at the QSO, located at the 0' origin.

We now discuss the more formal, algebraic properties of the SLT. It is convenient to measure time in units of length, defining a new time $\tau \equiv ct$. We then write the SLT (4.1) as

$$x' = \Gamma(x-\beta\tau), \quad \tau' = \Gamma(\tau-\beta x), \tag{4.5}$$

and can use matrix notation $X' = \Lambda X$, where $X \equiv (x,y,z,\tau)$ and is a column vector, and Λ is the 4 x 4 matrix

$$\Lambda = \begin{Bmatrix} \Gamma & 0 & 0 & -\beta\Gamma \\ 0 & 1 & 0 & 0 \\ 0 & 0 & 1 & 0 \\ -\beta\Gamma & 0 & 0 & \Gamma \end{Bmatrix} \tag{4.6}$$

The matrix $\Lambda = \Lambda(\beta)$ has an inverse $\Lambda(\beta)^{-1}$ such that $X = \Lambda^{-1}X'$ is the inverse transformation, and the reader can easily verify that $\Lambda(\beta)^{-1} = \Lambda(-\beta)$, and therefore

$$x = \Gamma(x' + \beta\tau'), \quad \tau = \Gamma(\tau' + \beta x'). \tag{4.7}$$

Also, we can do two SLT's in succession, with respective velocities β_1 , β_2 . Then $X'' = \Lambda(\beta_2)\,\Lambda(\beta_1)X = \Lambda(\beta_{12})X$, where one can show that $\beta_{12} = (\beta_1+\beta_2)/(1+\beta_1\beta_2)$. The reader should verify this. Thus the SLT's form what is said to be a mathematical group: The inverse of an SLT exists and is itself an SLT, and the product of two SLT's is a SLT. We showed in the previous section that the general LT is composed of two rotations, an SLT, and possibly inversions. Thus the general LT's also form a group.

It is convenient to consider the coordinate t to be the "fourth dimension" of space-time, and use the notation $\tau \equiv x^4$. We may then let $x \equiv x^1$, $y \equiv x^2$, $z \equiv x^3$, and write equations (4.1) as

$$x'^a = \Sigma_{b=1}^{4} \Lambda^a{}_b x^b, \quad a = 1, 2, 3, 4. \tag{4.8}$$

In general, we always use Latin indices to indicate the range 1,...,4. Greek indices will, by convention, here indicate the values 1, 2, 3. Many textbooks (for example, Rindler) use the opposite convention.

In order to avoid the burden of using summation signs too frequently, the <u>Einstein Summation Convention</u> is used, whereby a <u>subscript-superscript index pair in the same term using the same letter implies a summation over that pair</u>. Thus, $\Sigma_{a=1}^{4} A_a B^a \equiv A_a B^a \equiv A_1 B^1 + A_2 B^2 + A_3 B^3 + A_4 B^4$. Equation (4.8) would then be written $x'^a = \Lambda^a{}_1 x^1 + \Lambda^a{}_2 x^2 + \Lambda^a{}_3 x^3 + \Lambda^a{}_4 x^4 \equiv \Lambda^a{}_b x^b$. This convention is very useful in both special and general relativity and is in wide use in tensor analysis generally. Note that since a summation index is a dummy index it may be replaced by any letter. For example, $A_a B^a \equiv A_b B^b \equiv A_c B^c$, etc.

Another notational convenience is useful; namely the symbolic use of the character x for the whole set (x,y,z,τ). For example, the functional dependence of a function f on all four variables x, y, z, τ is often indicated by writing $f(x, y, z, \tau) \equiv f(x)$. Also in use is the notation $f(\vec{x}, \tau)$, which contrasts the spatial and temporal dependences; also, one may use $f(x^a)$ to mean the same thing. Thus, $f(x^a)$ usually does not mean dependence on only one of the variables.

We now introduce the definition of a form-invariant. Any function $\phi(x, y, x, \tau) \equiv \phi(x)$ which, under an LT, has the same functional dependence on the x'^a that it did on the x^a is called a form-invariant under an LT.

Consider the function $s^2 \equiv \tau^2 - x^2 - y^2 - z^2$. We show that s^2 is a form-invariant under a general homogeneous LT. Now, s^2 is obviously form-invariant under spatial rotations, since $x^2 + y^2 + z^2$ is simply the spatial distance from the origin. Thus we must only show that s^2 is form-invariant under an SLT. Then we have, from equations (4.1) and (4.7), $s^2 \equiv \tau^2 - x^2 - y^2 - z^2 = \Gamma^2(\tau' + \beta x')^2 - \Gamma^2(x' + \beta\tau')^2 + y'^2 + z'^2$, from which one can easily show that $s^2 = \tau'^2 - x'^2 - y'^2 - z'^2$. This has the same functional dependence on x'^a as it did on x^a, and thus we say that s^2 is form-invariant under homogeneous LT's.

Now consider the coordinate differentials dx, dy, dz, $d\tau$ between two infinitesimally separated neighboring events $P(x,y,z,\tau)$ and $P(x+dx, y+dy, z+dz, \tau+d\tau)$. We can construct the quadratic form

$$(ds)^2 \equiv -(dx)^2 - (dy)^2 - (dz)^2 + (d\tau)^2 \equiv \eta_{ab} dx^a dx^b , \tag{4.9}$$

with the summation convention implied over a and b, where η is the matrix

$$\eta \equiv \begin{Bmatrix} -1 & 0 & 0 & 0 \\ 0 & -1 & 0 & 0 \\ 0 & 0 & -1 & 0 \\ 0 & 0 & 0 & +1 \end{Bmatrix} \equiv (\eta_{ab}) . \tag{4.10}$$

Please note that ds is not meant to be the differential of the s defined by $s^2 \equiv \tau^2 - x^2 - y^2 - z^2$. This is confusing, but s^2 is not used very much in relativity, and the ds defined by equation (4.9) is the more fundamental, as we will see.

What is ds, physically speaking? Suppose we have a clock traversing a path in space-time so that it passes through the two infinitesimally separated events. Then the velocity of the clock with respect to the given SICS is given by $\beta^2 = (dx/d\tau)^2 + (dy/d\tau)^2 + (dz/d\tau)^2$, and thus $(ds)^2 = (d\tau)^2 (1-\beta^2)$, or $ds = d\tau (1-\beta^2)^{1/2}$, $d\tau = \gamma ds$, where $\gamma \equiv (1-\beta^2)^{-1/2}$. But this is just the time dilatation formula (4.4), with ds replacing $\Delta t'$. The clock reading t' in the time-dilatation discussion was moving with velocity β with respect to the SICS 0. This ds, defined by equation (4.9), is therefore, except for a factor of c, the time interval registered by a standard clock which passes through the two events. It is called the proper time interval between the two infinitesimally separated events.

The differential quadratic form given by equation (4.9) will be indispensible later on for treating the dynamics of particles undergoing accelerated motion. ds will then be the proper time required by a particle to go from the event $P(x,y,z,\tau)$ to the event $P(x+dx, y+dy, z+dz, \tau+d\tau)$, and one can show that ds^2 is form-invariant under LT's by taking the differentials of equations (4.7), $dx = \Gamma(dx' + \beta d\tau')$, $d\tau = \Gamma(d\tau' + \beta dx')$, and obtaining $ds^2 = -(dx')^2 - (dy')^2 - (dz')^2 + (d\tau')^2 = \eta_{ab} dx'^a dx'^b$.

Note that if a light beam in free space can pass directly from one event to the neighboring event, then ds = 0. The coordinate differential dx^a is then said to be null, or light-like. If $d\tau$ is so small that $ds^2 < 0$ (ds is imaginary) then it would be impossible for a material particle to pass from one event to the other, and dx^a is then said to be space-like. If $ds^2 > 0$, then dx^a is time-like.

We mention in passing that ds^2 is the central geometrical object of the general theory of relativity, where one writes

$$ds^2 = g_{ab} dx^a dx^b, \qquad (4.11)$$

with g_{ab} no longer given by equation (4.10). g_{ab} is called the metric tensor, and η_{ab} the Minkowski (or Lorentz) metric tensor. ds^2 is also called the metric.

Experiments on elementary particles decaying in flight have verified many times that ds is the proper time and that the time dilatation formula (4.4) is correct. Mesons moving at velocity β have been shown to have longer decay times than mesons at rest, in numerical agreement with equation (4.4). For further information and references on this point, the reader is referred to Schwartz, Section 3-3.

In special relativity, the differentials dx^a can be arbitrarily large, provided that the Lorentz metric η_{ab} is used, and provided that the particle or clock going between the two events moves like a free particle (in a straight line with uniform velocity). Then we write

$$(\Delta s)^2 = \eta_{ab}\Delta x^a \Delta x^b. \tag{4.12}$$

It might be useful to recall again our example of the QSO outbursts in Section (4.2). The coordinate time interval in the O' system between the two outbursts is given as 1 year $\simeq$ 3.16 x 10^7 sec, or $\Delta x'^4 = c\Delta t' = 9.48 \times 10^{17}$ cm, or one light year. Since the spatial intervals in the O' system are all zero, we have $\Delta s = \Delta x'^4 \simeq 9.48 \times 10^{17}$ cm. We now compute Δs in the O system and verify that the same answer is obtained. We have $\Delta x^4 = c\Delta t = 1.43 \times 10^{18}$cm. Then $\Delta x^1 = x_2 - x_1 = V\Delta t = 3/4\ c\Delta t = 1.07 \times 10^{18}$ cm, and so, from equation (4.11), $\Delta s = [(\Delta x^4)^2 - (\Delta x^1)^2]^{1/2} = [(1.43)^2 - (1.07)^2]^{1/2} \times 10^{18}$ cm $\simeq 9.45 \times 10^{17}$ cm, in good numerical agreement.

4. Vectors and Tensors in Relativity

The topic of tensor analysis is difficult for most students to broach. Vectors are certainly easy to picture, but tensors seem much more abstract and hard to visualize, particularly in the four dimensions of relativity theory. The fact is, however, that tensors are best treated mathematically, and from an abstract point of view. This is no real hindrance, for their rules of manipulation are really quite simple and esthetically satisfying, once the terrors of abstraction are banished by a little familiarization.

The reason for studying tensors in relativity theory is that they are a natural vehicle for expressing physical laws. The postulated equivalence of SICS's in special relativity means that the basic equations of physics (for example, Maxwell's equations and the Lorentz force equation) must not change their form when expressed in terms of the coordinates of different SICS's. Tensor equations will be seen to have exactly this property. Thus the basic quantities of physics, such as momentum, energy, and the electromagnetic field components, will be used to construct tensors.

The first "tensor" that we consider is a rather trivial one, and is called a scalar, or an invariant. Any function f(x) of space-time which does not change its value under coordinate transformations is called a scalar (invariant). By this we mean that if we change coordinates in some arbitrary, possibly non-linear, fashion by means of the transformation $x^a = x^a(\bar{x})$, then in the barred system, $\bar{f}(\bar{x}) = f(x(\bar{x}))$. In this treatment we do not restrict ourselves to LT's, which are linear, since the non-linear case is no more trouble to do, and since the tensor analysis thus learned can be used also for the general theory of relativity.

Any quantity which has a coordinate-free definition is a scalar. For example, the proper-time interval ds is a scalar, since it is defined as the time read from a standard clock passing between the two neighboring events. It is perhaps somewhat more clear to consider a clock moving along a general, accelerated space-time path parameterized by a parameter λ, so that $x^a = x^a(\lambda)$. Then the running (proper) time read by the clock is $s=s(\lambda)$. Under a coordinate transformation $\bar{x}^b = \bar{x}^b(x)$, we would have $\bar{x}^b(\lambda) = \bar{x}^b(x(\lambda))$, and $\bar{s}=s=s(\lambda)$ still. This is quite trivial, and the basic idea should be clear.

A scalar is called a tensor of zero rank.

We now consider vectors, which are tensors of the first rank. There are two types of vectors, contravariant and covariant, and we define contravariant vectors first. Suppose we have a set of four space-time functions X^a which are defined in the coordinate system x^a, and suppose that in the new coordinate system $\bar{x}^a$ these quantities are $\bar{X}^a$. Then if for all transformations $\bar{x}^a = \bar{x}^a(x)$ the X^a and $\bar{X}^a$ are related by the formula

$$\bar{X}^a = \frac{\partial \bar{x}^a}{\partial x^b} X^b , \tag{4.13}$$

with summation over b implied, then X^a (and $\bar{X}^a$) is a contravariant vector, or contravariant first-rank tensor. The summation convention is here broadened a bit to include superscript pairs if one of the superscripts occurs in the denominator of a partial derivative.

It is easy to find an example of a contravariant vector. From advanced calculus, we know that a coordinate differential $dx^n = (dx^1, dx^2, dx^3, dx^4)$ transforms like $d\overline{x}^n = \frac{\partial \overline{x}^n}{\partial x^m} dx^m$. (Remember that dx^m is the set of coordinate differences between a pair of neighboring events). Thus dx^a is a contravariant vector. This can be made somewhat more physical by considering the motion of a particle, where dx^a is an increment of the particle's coordinates. (The two neighboring events are defined by the particle's successive passage through two neighboring points in 3-space.) The particle's "ordinary" velocity is dx^α/dx^4 $(= v^\alpha/c)$, which has no obvious four-dimensional tensor properties. However, if s is the proper time associated with the particle (read from a standard clock moving with the particle), then the quantity $U^a \equiv dx^a/ds$ is a contravariant vector, since ds is a scalar. This vector is called the <u>four-velocity</u> of the particle and will be very useful to us in the Section 6 in finding the relativistic analog of Newton's third law.

The statement (4.13) means that each of the quantities involved must be evaluated at the same event when the equality is tested, i.e., the equation $\overline{X}^a = \overline{X}^a(\overline{x}) = \overline{X}^a(\overline{x}(x)) = \frac{\partial \overline{x}^a(x)}{\partial x^b} X^b(x)$ must be an identity in x^b. It is simple to construct vectors by letting X^a be any four functions of x^a, and then defining their values in new systems $\overline{x}^a$ by means of (4.13). From a mathematical standpoint this is quite legitimate, but is not very interesting physically.

We now define a covariant vector in a similar way. A set of functions B_a which transform like

$$\overline{B}_a = \frac{\partial x^b}{\partial \overline{x}^a} B_b \tag{4.14}$$

is said to be a <u>covariant vector</u>, or <u>covariant first-rank tensor</u>.

Again, it is easy to find a familiar example. Let f(x) be a scalar function. Then we know from advanced calculus that the gradient $\partial f/\partial x^a$ transforms like $\frac{\partial \bar{f}}{\partial \bar{x}^a} = \frac{\partial x^b}{\partial \bar{x}^a} \frac{\partial f}{\partial x^b}$. Thus $\partial f/\partial x^a$ is a covariant vector.

As a simple illustration, let us consider the two-dimensional linear transformation $\bar{z}^1 = Az^1 + Bz^2$, $\bar{z}^2 = Cz^1 + Dz^2$, where A, B, C, and D are constants. Then equation (4.13), with $X^b = dz^b$, is written $d\bar{z}^1 = Adz^1 + Bdz^2$, $d\bar{z}^2 = Cdz^1 + Ddz^2$. To use equation (4.14), we must solve the transformation equation for z^a to get $z^1 = \mathcal{D}(D\bar{z}^1 - B\bar{z}^2)$ and $z^2 = \mathcal{D}(A\bar{z}^2 - C\bar{z}^1)$, where $\mathcal{D} \equiv (AD - BC)^{-1}$. Then $\partial f/\partial\bar{z}^1 = \mathcal{D}(D\partial f/\partial z^1 - C\partial f/dz^2)$, and $\partial f/\partial \bar{z}^2 = \mathcal{D}(-B\partial f/\partial z^1 + A\partial f/\partial z^2)$. Note that dz^a and $\partial f/\partial z^a$ transform in very different ways. This must be the case, since the differential $df = (\partial f/\partial z^a)dz^a$ is a scalar invariant, whose value is independent of the coordinate system, since it is the difference between the invariant values of f at two neighboring fixed points $(z^a, z^a + dz^a)$. The reader should verify by direct substitution that $df = (\partial f/\partial \bar{z}^1)d\bar{z}^1 + (\partial f/\partial \bar{z}^2)d\bar{z}^2 = (\partial f/\partial z^1)dz^1 + (\partial f/\partial z^2)dz^2$.

For the SLT, we set $x = z^1$ and $\tau = z^2$, and use equations (4.5) and (4.7) to get $dx' = \Gamma(dx - \beta d\tau)$, $d\tau' = \Gamma(d\tau - \beta dx)$, as in Section (4.3) and $\partial f/\partial x' = \Gamma(\partial f/\partial x + \beta\partial f/\partial\tau)$, $\partial f/\partial\tau' = \Gamma(\partial f/\partial\tau + \beta\partial f/\partial x)$.

The definitions of covariance and contravariance are easily extended to tensors of rank greater than first. Suppose we have a set of 16 quantities T^{ab} Then T^{ab} is a <u>second-rank contravariant tensor</u> if it transforms like

$$\bar{T}^{ab} = \frac{\partial \bar{x}^a}{\partial x^m} \frac{\partial \bar{x}^b}{\partial x^n} T^{mn}. \tag{4.15}$$

Similarly, S_{ab} is a <u>covariant second-rank tensor</u> if

$$\bar{S}_{ab} = \frac{\partial x^m}{\partial \bar{x}^a} \frac{\partial x^n}{\partial \bar{x}^b} S_{mn}. \tag{4.16}$$

The generalization to higher-order tensors is obvious.

One may also speak of tensors of mixed character. The quantities $T^a{}_b$ form a mixed second-rank tensor if

$$\bar{T}^m{}_n = \frac{\partial \bar{x}^m}{\partial x^a} \frac{\partial x^b}{\partial \bar{x}^n} T^a{}_b \, . \tag{4.17}$$

To construct an example of a second-rank tensor, let A^a and B^b be contravariant vectors. Then the 16 products $A^a B^b$ constitute a second-rank contravariant vector, as the reader may easily verify. In Section (4.5), we show that the electromagnetic field may be represented as a second-rank tensor F_{ab} .

We should mention here the important point that the sum or difference of two tensors of the same type is itself a tensor, as the reader may verify. Also, if in a particular coordinate system all components of a tensor vanish, then they vanish in all coordinate systems. It follows then that if two tensors are component-wise equal in one coordinate system, they are equal in all others. This shows the great utility of tensor equations for expressing physical laws in both special and general relativity.

Let us for the moment restrict ourselves to LT's and investigate the algebraic properties of the Minkowski metric tensor η_{ab}. The matrix Λ, given by equation (4.6) is of course equal to $\partial x'^a/\partial x^b$, so that $\Lambda^a{}_b = \partial x'^a/\partial x^b$. Then the coordinate differentials transform like $dx'^a = \Lambda^a{}_b dx^b$. We know already that the Minkowski metric ds^2 (equation (4.9)) is form-invariant under LT's and thus we can write $ds^2 = \eta_{ab}dx'^a dx'^b = \eta_{ab}\Lambda^a{}_m\Lambda^b{}_n \, dx^m dx^n = \eta_{mn}dx^m dx^n$, with independent summations implied over a,b,m, and n. But this is true for all dx^m, and therefore

$$\eta'_{mn} = \eta_{mn} = \Lambda^a{}_m \Lambda^b{}_n \eta_{ab} \, . \tag{4.18}$$

Thus we may say that η is a second-rank covariant tensor under an LT; and further, it is form-invariant, in the sense that the new η'_{mn} is identical to the old η_{ab}. The reader is advised to verify equation (4.18) directly from equations (4.6) and (4.10).

It is instructive to consider what happens to the general metric form $ds^2 = g_{ab}\, dx^a dx^b$ under coordinate transformations that are not LT's. Since ds is, by definition, the differential proper time between neighboring events, its value does not depend on the coordinate system in use. We have stated this before in noting that ds is a scalar. Therefore, by equation (4.13) with the barred and non-barred coordinates interchanged, $ds^2 = g_{ab}\, dx^a dx^b = g_{ab} \frac{\partial x^a}{\partial \bar{x}^m} \frac{\partial x^b}{\partial \bar{x}^n} d\bar{x}^m d\bar{x}^n = \bar{g}_{mn}\, d\bar{x}^m\, d\bar{x}^n$. Since these equations are true for all $d\bar{x}^a$, one can show that

$$\bar{g}_{mn} + \bar{g}_{nm} = \left(\frac{\partial x^a}{\partial \bar{x}^m} \frac{\partial x^b}{\partial \bar{x}^n} + \frac{\partial x^a}{\partial \bar{x}^n} \frac{\partial x^b}{\partial \bar{x}^m} \right) g_{ab} \,. \tag{4.19}$$

Now, since the symmetric combination $g_{ab} + g_{ba}$ is all that really matters in equation (4.11), g_{ab} is always chosen to be symmetric in relativity; i.e., $g_{ab} = g_{ba}$. This is certainly true for η_{ab}, which is diagonal. Theories in which the metric tensor is not symmetric have been constructed, but they do not seem to have led to anything fruitful. For the symmetric case, equation (4.19) reduces to

$$\bar{g}_{mn} = \frac{\partial x^a}{\partial \bar{x}^m} \frac{\partial x^b}{\partial \bar{x}^n} g_{ab} \,. \tag{4.20}$$

Therefore, g_{ab} is a second-rank covariant tensor. In general relativity, it is not possible to set up coordinates so that g_{ab} becomes the Minkowski metric tensor everywhere in space-time.

The reader should prove equation (4.19) by substituting various forms of $d\bar{x}^a$. First $d\bar{x}^a = (dx, 0,0,0)$, etc., then (dx, dy, 0,0), etc., ad nauseam. One should also show that equation (4.20) follows from equation (4.19) if g_{ab} is symmetric, by exchanging summation indices in equation (4.19). Also, the reader should prove that symmetry of tensor indices is invariant under a general coordinate transform; i.e., if $g_{ab} = g_{ba}$ is a tensor, then $\bar{g}_{ab} = \bar{g}_{ba}$.

There are also antisymmetric tensors, satisfying $F_{ab} = -F_{ba}$. In this case the diagonal components are all zero; for example, for a = b = 1, $F_{11} = -F_{11}$. Antisymmetry is also invariant under a coordinate transformation, as the reader should prove for himself.

At this point, it may be helpful to pause and to recapitulate what we have done as far as the metric tensors g_{ab} and η_{ab} and the related proper time ds are concerned.

By using Einstein's postulated relativity principle, together with the reciprocity and homogeneity and isotropy hypotheses, one derives the LT. We were then able to show that the Minkowski metric is form-invariant under the LT and is the proper time between infinitesimally separated events. Thus the assertion that ds is a proper time interval follows from Einstein's relativity principle and the reciprocity and homogeneity and isotropy hypotheses. This assertion has been verified by experiment, as we have noted, as has the isotropy principle, by means of the Michelson-Morley experiment. For the details of this experiment, see for example Schwartz, Section 2-2 (Problems), and Panofsky and Phillips, Chapter 14.

We have also proved the mathematical fact that the Minkowski metric tensor η_{ab} is a form-invariant second-rank tensor under LT's. This is sheer algebra and is independent of the interpretation of ds as proper time. It is, however, equivalent to the mathematical fact that $ds^2 = d\tau^2 - dx^2 - dy^2 - dz^2$ is form-invariant under LT's.

We then generalized the metric tensor to embrace the metric form $ds^2 = g_{ab}dx^a dx^b$ for arbitrary space-time coordinates. The continued requirement that ds be an invariant proper time then implies that g_{ab} is a second-rank covariant tensor. This is, again, a mathematical fact, but it can be verified indirectly by showing experimentally that the ds so constructed is indeed the proper time. One can go further than this and exhibit a procedure for measuring the individual components of g_{ab}, for a given coordinate system (Landau and Lifshitz, 1958, Section 84).

As a simple example, we show how to derive the metric tensor for cylindrical polar coordinates in special relativity. The time coordinate will not be transformed at all.

Let $x = r\cos\phi$, $y = r\sin\phi$, with $\overline{x^1} = r$, $\overline{x^2} = \phi$, $\overline{x^3} = z$, $\overline{x^4} = \tau$. Then since $g_{ab} = \eta_{ab}$, equation (4.20) gives

$$\bar{g}_{11} = \frac{\partial x^a}{\partial r}\frac{\partial x^b}{\partial r}\eta_{ab} = -\left(\frac{\partial x}{\partial r}\right)^2 - \left(\frac{\partial y}{\partial r}\right)^2 = -\cos^2\phi - \sin^2\phi = -1$$

$$\bar{g}_{12} = \frac{\partial x^a}{\partial r}\frac{\partial x^b}{\partial \phi}\eta_{ab} = -\frac{\partial x}{\partial r}\frac{\partial x}{\partial \phi} - \frac{\partial y}{\partial r}\frac{\partial y}{\partial \phi} = 0$$

$$\bar{g}_{22} = \frac{\partial x^a}{\partial \phi}\frac{\partial x^b}{\partial \phi}\eta_{ab} = -\left(\frac{\partial x}{\partial \phi}\right)^2 - \left(\frac{\partial y}{\partial \phi}\right)^2 = -r^2, \text{ etc.}$$

The final result is $ds^2 = -dr^2 - r^2 d\phi^2 - dz^2 + d\tau^2$.

A physically more interesting coordinate transformation is to one in which the coordinates are rotating about the z axis at some angular velocity Ω. Let us now write $x^1 = r$, $x^2 = \phi$, $x^3 = z$, $x^4 = \tau$, and let $x^1 = \bar{x}^1$, $x^2 = \bar{x}^2 + \frac{\Omega}{c}\bar{x}^4$, $x^3 = \bar{x}^3$, $x^4 = \bar{x}^4$. One can then verify that $\bar{g}_{11} = g_{11} = -1$, $\bar{g}_{22} = g_{22} = -r^2$, $\bar{g}_{33} = g_{33} = -1$, $\bar{g}_{44} = (\partial x^2/\partial\bar{x}^4)^2 g_{22} + (\partial x^4/\partial\bar{x}^4)^2 g_{44} = 1 - \Omega^2 r^2/c^2$, $\bar{g}_{24} = \bar{g}_{42} = -\Omega r^2/c$. Thus we have, calling $\bar{\phi} \equiv \bar{x}^2$, $ds^2 = -dr^2 - r^2 d\bar{\phi}^2 - dz^2 - 2\Omega r^2 d\bar{\phi}\,d\tau/c + (1 - \Omega^2 r^2/c^2)d\tau^2$.

Note the appearance of the cross-term in $d\bar{\phi}\,d\tau$. Note also that for $r > c/\Omega$, we have $\bar{g}_{44} < 0$. Thus for a particle trajectory satisfying $dr = d\bar{\phi} = dz = 0$ (i.e., for a particle at rest with respect to the rotating coordinate system) at $r > c/\Omega$ the proper time is imaginary! But such a particle would be moving at a speed faster than the speed of light relative to the original non-rotating coordinate system, and thus we see that our tensor analysis has certain built-in safeguards.

5. The Electromagnetic Field and Maxwell's Equations

We now proceed to establish certain laws of physics in tensor form. The general philosophy of this approach will be to make reasonable assumptions about the tensor character of certain physical qualities, and then to see if existing non-relativistic equations can be suitably generalized into relatavistic tensor form. This approach was the one actually used historically by the original investigators, who arrived by trial and error at the tensor forms used today. Of course, the results are subject to verification by experiment.

Let us first consider the electromagnetic field and Maxwell's equations. The pair of Maxwell's equations which do not involve the charge and current are

$$\vec{\nabla}\cdot\vec{B} = 0 \tag{4.21a}$$

$$\vec{\nabla}\times\vec{E} = -\frac{\partial\vec{B}}{\partial\tau} \ . \tag{4.21b}$$

Apparently, $\vec{E}$ and $\vec{B}$ might be considered to be different components of the same tensor, since they are both present together in equation (4.21b). This is verified by the experimental fact that $\vec{E}$ and $\vec{B}$ do not transform independently under an LT. For example, a point charge at rest in one SICS generates only an electric field. Seen from the standpoint of a moving frame, however, the point charge appears as a current as well, and thus a magnetic field arises. Now, $\vec{E}$ and $\vec{B}$ together have six independent components, and hence could not together make a relativistic vector, which has only four components. However, an antisymmetric tensor F_{ab} does have six independent components. The reader should refer back to the end of the previous section for a discussion of the symmetry properties of tensors. Accordingly, let us try the form

$$(F_{ab}) = \begin{pmatrix} 0 & B_z & -B_y & E_x \\ -B_z & 0 & B_x & E_y \\ B_y & -B_x & 0 & E_z \\ -E_x & -E_y & -E_z & 0 \end{pmatrix} = -(F_{ba}), \qquad (4.22)$$

where the left-hand index denumerates the rows and the right-hand one the columns. Again, this particular arrangement is used because it provides us with tensor equations, as we will see.

Let us now introduce notation which will simplify our typography a bit. We denote the partial derivative $\partial/\partial x^a$ by a comma followed by a subscript $_a$, and write $\partial f/\partial x^a \equiv f,_a$. Then equation (4.21a) becomes, using equation (4.22), $\vec{\nabla}\cdot\vec{B} = B_{x,1} + B_{y,2} + B_{z,3} = F_{23,1} + F_{31,2} + F_{12,3} = 0$.

The x component of equation (4.21b) may be written $(\vec{\nabla}X\vec{E})_x + \partial B_x/\partial\tau = E_{z,2} - E_{y,3} + B_{x,4} = F_{34,2} + F_{42,3} + F_{23,4} = 0$. This suggests that equations (4.21) may be replaced by the set of equations

$$F_{ab,c} + F_{ca,b} + F_{bc,a} = 0 . \qquad (4.23)$$

The reader should verify that the y and z components of equations (4.21b) conform to this relation, and that this relation contains no other equations than (4.21).

But is an expression like $F_{ab,c}$ a tensor? We know already that $f,_a$ is a first-rank tensor if f is a scalar. Differentiating equation (4.16), we find

$\overline{F}_{ab,c} = \frac{\partial\overline{F}_{ab}}{\partial\overline{x}^c} = \frac{\partial x^n}{\partial\overline{x}^a}\frac{\partial x^m}{\partial\overline{x}^b}\frac{\partial F_{nm}}{\partial\overline{x}^c} + F_{nm}\frac{\partial}{\partial\overline{x}^c}\left(\frac{\partial x^n}{\partial\overline{x}^a}\frac{\partial x^m}{\partial\overline{x}^b}\right)$. But $\frac{\partial F_{nm}}{\partial\overline{x}^c} = \frac{\partial x^p}{\partial\overline{x}^c}\frac{\partial F_{nm}}{\partial x^p}$, and hence

$$\overline{F}_{ab,c} = \frac{\partial x^n}{\partial\overline{x}^a}\frac{\partial x^m}{\partial\overline{x}^b}\frac{\partial x^p}{\partial\overline{x}^c} F_{nm,p} + F_{nm}\frac{\partial}{\partial\overline{x}^c}\left(\frac{\partial x^n}{\partial\overline{x}^a}\frac{\partial x^m}{\partial\overline{x}^b}\right) . \qquad (4.24)$$

Therefore $F_{ab,c}$ is a (third-rank) tensor only if the term on the far right vanishes. But it does if we consider only LT's, which are linear! Thus $F_{ab,c}$ is a tensor for linear transformations. The generalization of the derivative so that it is tensorial under non-linear transformations leads to what is known as covariant differentiation and is beyond the scope of the present work. As it turns out, however, one can use equation (4.24) to show that the expression $\Phi_{abc} \equiv F_{ab,c} + F_{ca,b} + F_{bc,a}$ is as a whole a tensor, provided that $F_{ab} = - F_{ba}$. Thus if $\Phi_{abc} = 0$ (equation 4.23) in one coordinate system, it is also true in any other.

Let us note that the hypothesis that the electromagnetic field forms a tensor in the manner of equation (4.22) implies that the fields must transform under an LT in accordance with equation (4.16), with Λ replacing the partial derivatives. For the SLT one can readily show that the transformed fields are given by

$$\begin{aligned} E_x' &= E_x & B_x' &= B_x \\ E_y' &= \Gamma(E_y - \beta B_z) & B_y' &= \Gamma(B_y + \beta E_z) \\ E_z' &= \Gamma(E_z + \beta B_y) & B_z' &= \Gamma(B_z - \beta E_y). \end{aligned} \tag{4.25}$$

No experimental evidence exists which contradicts these relations.

As a useful application, we consider the field outside a uniform wire (or infinitely long cylinder) of charge. In the SICS with respect to which the charge is at rest, it is well known that the electric field strength outside the wire is $E=2\sigma/r$, where r is the distance to the wire and σ is the charge per unit length (esu/cm). We choose the x-axis to be along the wire, so that $E_x = 0$ and $E = (E_y^2 + E_z^2)^{1/2}$. If we perform an SLT along the direction of the wire, in the O' system the moving charge will appear as a current. Equations (4.25) then predict what the magnetic field in the O' system will be. Since $B = 0$, we have immediately $B_x' = B_x = 0$, $B_y' = \Gamma\beta E_z$, and $B_z' = -\Gamma\beta E_y$, so that $B' = (B_y'^2 + B_z'^2)^{1/2} = \Gamma\beta(E_y^2 + E_z^2)^{1/2} = 2\sigma\Gamma\beta/r$. The distance r does not change under the SLT since it is a length perpendicular to the direction of motion. Therefore, equations (4.25) predict a result in complete agreement with the usual

$B = 2I/cr$ (Jackson, pg. 135), where I is the current carried by the wire, provided that we take $I = \sigma\Gamma V = \sigma' V$, where $\sigma' = \sigma\Gamma$ is the charge density cm^{-1} observed in O'. This is very reasonable, since the wire has suffered a Lorentz contraction, so that in the O' system we see more esu/cm than in the O system. The question of how charges and currents transform generally is treated at the end of this section.

Before treating the second pair of Maxwell's equations, it is necessary to do a little more tensor analysis and to introduce the operations of <u>contracting</u>, and of <u>raising</u> and <u>lowering</u> tensor indices. It will be necessary to employ the following identities from advanced calculus, which hold for any coordinate transformation:

$$\frac{\partial \bar{x}^k}{\partial \bar{x}^m} = \frac{\partial x^\ell}{\partial \bar{x}^m}\frac{\partial \bar{x}^k}{\partial x^\ell} = \delta^k{}_m, \quad \frac{\partial x^a}{\partial x^b} = \frac{\partial \bar{x}^c}{\partial x^b}\frac{\partial x^a}{\partial \bar{x}^c} = \delta^a{}_b, \qquad (4.26)$$

where $\delta^a{}_b$ is the Kronecker delta ($\delta^a{}_b = 1$ for $a = b$, $\delta^a{}_b = 0$ for $a \neq b$).

For the SLT, equation (4.6), this is stated as $\Lambda(\beta)\,\Lambda(-\beta) = I$, or $\Lambda(\beta)^{-1} = \Lambda(-\beta)$, as the reader should verify.

The operation of <u>contraction</u> is very simple, and has been used before; e.g., in equation (4.9). It is simply a summation over a pair of covariant and contravariant indices. For example, if A^a and B_a are contravariant and covariant vectors, then $A^a B_b$ is a mixed tensor, and its contraction consists simply in setting $a = b$ and summing in accordance with the summation convention: $A^a B_a = A^1 B_1 + A^2 B_2 + A^3 B_3 + A^4 B_4$. It is important to note that a scalar is thereby created, since we have $\bar{A}^m \bar{B}_m = A^a \frac{\partial \bar{x}^m}{\partial x^a}\frac{\partial x^b}{\partial \bar{x}^m} B_b = A^a \delta^b{}_a B_b = A^a B_a$, from equation (4.26). This is known as the scalar product of A and B.

The contraction operation may be used for any mixed tensor expression. For example, consider the third-rank tensor $T^a{}_{bc}$. The contractions $T^a{}_{ba} = T^1{}_{b1} + T^2{}_{b2} + T^3{}_{b3} + T^4{}_{b4}$ and $T^a{}_{ac}$ are both covariant vectors, as the reader should prove for himself using equation (4.26).

The metric tensor g_{ab}, or its Minkowski counterpart η_{ab}, is often contracted with contravariant indices to produce covariant ones. If A^a is a contravariant vector, the expression $g_{ab}A^b \equiv A_a$ is a covariant one. This is called <u>lowering</u> an index.

Now consider the matrix inverse g^{-1} of the metric tensor g, which exists if the determinant $|g| \neq 0$. This inverse is denoted by g^{ab}, and satisfies the relation $g^{ab}g_{bc} = \delta^a{}_c$. One can easily show that $\delta^a{}_c$ is a form-invariant mixed tensor, and that g^{ab} is a contravariant tensor. Then g^{ab} may be used to raise indices. If S_{ab} is a covariant tensor, then $S^a{}_b \equiv g^{ac} S_{cb}$ is a mixed tensor, and $S^{ab} = g^{ac}g^{bd}S_{cd}$ is a contravariant tensor. Raised indices may be consistently lowered again, and vice versa, since, for example, $A^a = g^{ab}A_b = g^{ab}g_{bc}A^c = \delta^a{}_cA^c = A^a$.

For the Minkowski metric, these operations are somewhat trivial and involve only changes of sign. It is interesting in this connection that η is its own inverse; i.e., $\eta = \eta^{-1}$, or $\eta^2 = I$, as is readily apparent from equation (4.10). In this case raising or lowering a spatial index changes the sign of a component; e.g., $A_1 = \eta_{1a}A^a = \eta_{11} A^1 = -A^1$. For the temporal component, nothing happens: $A_4 = \eta_{4a}A^a = \eta_{44}A^4 = +A^4$. These sign changes seem trivial, but it is necessary to keep track of them in order to treat the Lorentz force equation and the rest of Maxwell's equations.

Let us continue with Maxwell's equations, the second pair of which are written, in Gaussian e.s.u.,

$$\vec{\nabla}\times\vec{B} - \frac{\partial\vec{E}}{\partial\tau} = \frac{4\pi}{c}\vec{j}\,, \tag{4.27a}$$

$$\vec{\nabla}\cdot\vec{E} = 4\pi\rho\,. \tag{4.27b}$$

Consider the expression $\vec{\nabla}\cdot\vec{E} = -F_{41,1} - F_{42,2} - F_{43,3}$ (see equation(4.22)). The occurrence of the index pairs 1, 1; 2, 2; 3, 3 prompts us to try to convert this into a tensor contraction. We can write, since raising a spatial index changes the sign, and since $F^{44} = 0$, $\vec{\nabla}\cdot\vec{E} = F^{41}{}_{,1} + F^{42}{}_{,2} + F^{43}{}_{,3} + F^{44}{}_{,4}$. Then similarly $(\vec{\nabla}\times\vec{B})_x - \partial E_x/\partial\tau = F_{12,2} + F_{13,3} - F_{14,4} = F^{11}{}_{,1} + F^{12}{}_{,2} + F^{13}{}_{,3} + F^{14}{}_{,4}$, and therefore we have $F^{\alpha a}{}_{,a} = 4\pi j^\alpha/c$ $(\alpha = 1, 2, 3)$, and $F^{4a}{}_{,a} = 4\pi\rho$. Now we already know that $F^{ab}{}_{,c}$ is a tensor under the LT, and therefore the contraction $F^{ab}{}_{,b}$ is a contravariant Lorentz vector. It should therefore be concluded that the current density $j^\alpha/c \equiv J^\alpha$ and the charge density $\rho \equiv J^4$ together also form a contravariant Lorentz vector. This conclusion is, of course, subject to verification by experiment. The vector J^a is known as the charge-current four-vector, and the second pair of Maxwell's equations is then

$$F^{ab}{}_{,b} = 4\pi J^a. \tag{4.28}$$

The reader should prove that the equation of conservation of charge is now identically satisfied if J^a is given by equation (4.28), where F^{ab} is antisymmetric. It is written

$$\vec{\nabla}\cdot\vec{j} + \frac{\partial\rho}{\partial t} = cJ^a{}_{,a} = 0 .$$

Knowing that $J^a = (\vec{j}c^{-1}, \rho)$ is a contravariant vector, we use the transformation law (4.13) to finish the discussion of the magnetic field about a moving wire of charge. The only thing left to verify is that the linear charge density σ(esu/cm) does indeed transform like $\sigma' = \Gamma\sigma$. Let R be the radius of the wire. Then the charge density ρ (esu/cm^3) is $\rho = \sigma(\pi R^2)^{-1}$. In the O' system we have $J'^a = \Lambda^a{}_b J^b$, where $J^b = (0,0,0,\rho)$, since the charge is at rest in O. Then $\rho' = J'^4 = \Lambda^4{}_b J^b = \Lambda^4{}_4 J^4 = \Gamma\rho$. But R is a length perpendicular to the transformation velocity, and hence does not change under an LT. Therefore, $\rho' = \sigma'(\pi R^2)^{-1}$, and hence $\rho' = \sigma'(\pi R^2)^{-1} = \Gamma\rho = \Gamma\sigma(\pi R^2)^{-1}$, or $\sigma' = \Gamma\sigma$. Also, for a = 1, we have $J'^1 = \Lambda^1{}_4 J^4 = -\beta\Gamma\rho$. Note the minus sign, which arises because the current appears to move in the negative x' direction (if $\sigma > 0$), as it should. But $j^\alpha = cJ^\alpha$ is the current density (esu/cm^{-2} sec^{-1}), and so the total current is $I = |\vec{j}|\,\pi R^2 = |J^\alpha|\, c\pi R^2 = \sigma\Gamma V$, in agreement with our previous conclusions.

6. The Dynamics of a Particle and the Lorentz Force Equation

We now wish to find the tensor form of the equation of motion of a particle, i.e., the relativistic replacement for Newton's law $\vec{f} = m\vec{a} = d\vec{p}/dt$. In Section (4.4), we noted that the quantity $U^a \equiv dx^a/ds$, the four-velocity, was the relativistic analog of the usual Newtonian velocity. Now, we found in Section (4.3) that $ds = d\tau/\gamma$, and therefore it follows that U^a may be written in component form as

$$(U^a) = c^{-1}\gamma(v_x, v_y, v_z, c) , \tag{4.29}$$

where $\gamma \equiv (1 - \beta^2)^{-1/2}$. The distinction between Γ for a Lorentz transformation and γ for a particle is made to avoid confusion.

The reader should verify the following identity, which follows from equation (4.29)

$$U_aU^a = \eta_{ab}U^aU^b = \frac{\eta_{ab}dx^a dx^b}{(ds)^2} = -(U^1)^2-(U^2)^2-(U^3)^2+(U^4)^2 = 1. \tag{4.30}$$

It is useful to check the non-relativistic limit of (4.29). As $\beta \to 0$, $U^4 = \gamma \to 1$, and $cU^\alpha \to v^\alpha$. Thus the spatial components of the vector cU^a go over into the usual Newtonian velocity in the non-relativistic limit.

We may also define the four-momentum of a particle. If m_o is the rest-mass of the particle we write the four-momentum as

$$P^a \equiv m_o c U^a . \tag{4.31}$$

From equation (4.30), we find

$$P^a P_a = m_o^2 c^2 . \tag{4.32}$$

What about acceleration? If dx^a/ds is the generalization of the velocity, then $d^2x^a/ds^2 = dU^a/ds$ should be the appropriate generalization of the acceleration. However, dU^a/ds is not a tensor. In fact, $\frac{d\bar{U}^b}{ds} = \frac{d}{ds}(\frac{\partial\bar{x}^b}{\partial x^a}U^a) = \frac{\partial\bar{x}^b}{\partial x^a}\frac{dU^a}{ds} + U^a \frac{dx^c}{ds}\frac{\partial^2\bar{x}^b}{\partial x^c \partial x^b}$, and therefore dU^a/ds is a contravariant vector only for LT's or other linear transformations. The true tensor form of the acceleration is properly treated in texts on general relativity. The appropriate replacement for $\vec{f}=m\vec{a}$ in special relativity is then

$$c\,\frac{dP^a}{ds} = m_o c^2\,\frac{dU^a}{ds} = f^a\,, \qquad (4.33)$$

where f^a is an appropriate "four-force".

Equation (4.33) is unfortunately devoid of content unless we know what f^a is. Let us as an example investigate the motion of a charged particle in an electromagnetic field, which is known to be

$$\frac{d\vec{p}}{dt} = e(\vec{E} + \frac{\vec{v}X\vec{B}}{c}). \qquad (4.34)$$

This is the Lorentz force equation, and although its form is non-tensorial, it is known experimentally to be valid for relativistic motion, provided $\vec{p} = m_o\gamma\vec{v}$, as per equations (4.29) and (4.31). Thus we must be able to convert equation (4.34) directly into Lorentz tensor form without any more guesswork. From equations (4.29) and (4.31) it follows that $dP^\alpha/dt = c(ds/d\tau)dP^\alpha/ds = c(U^4)^{-1}dP^\alpha/ds$. Let us study the x component of equation (4.34). Then we have from equations (4.22) and (4.29)

$$c(U^4)^{-1}\,\frac{dP^1}{ds} = -\,e(F_{41} + \frac{U^2}{U^4}F_{21} + \frac{U^3}{U^4}F_{31}).$$

After a little algebra, one finds $c\,\frac{dP^1}{ds} = eU^b F_b{}^1$, which tells us that the full tensor form of equation (4.34) should be

$$c\,\frac{dP^a}{ds} = eU^b F_b{}^a. \qquad (4.35)$$

One can easily show that the y and z components of equation (4.34) can be written in this form. But what is the a=4 component of equation (4.35)? One may use equations (4.22), (4.29), and (4.31) to show that a=4 implies the relation

$$\frac{d}{dt}(m_o c^2 \gamma) = \vec{f}\cdot\vec{v} \quad = e\vec{E}\cdot\vec{v} \ . \tag{4.36}$$

Now the right hand side of this equation is simply the rate at which the electromagnetic field is doing work on the particle. Therefore, the expression $m_o c^2 \gamma$ must be the kinetic energy of the particle, except for an additive constant. We note that, since $\vec{p} = m_o \gamma \vec{v}$ and $\vec{p} = m\vec{v}$ generally, we have $m = m_o \gamma$. The suggestion is, therefore, that $E = mc^2$, a famous conclusion of Einstein's.

The fourth component of the four-momentum is such that $cP^4 = m_o c^2 \gamma$ is said to be the total energy of the particle (except for potential energy), including a "rest energy" $m_o c^2$. The reader should check to see that $m_o c^2 \gamma$ may be expanded in a power series in β to obtain the non-relativistic limit $E \simeq m_o c^2 + (1/2)\, m_o c^2$. Thus equation (4.35) is compatible with Newtonian energy conservation when $v \ll c$.

It is easy to show that if $\hat{n} \equiv \vec{P}/P$ ($P \equiv |\vec{P}|$) is the unit 3-vector in the direction of the spatial part of the 4-momentum, then

$$P^a = P[\hat{n}, (1 + m_o{}^2 c^2 P^{-2})^{1/2}]. \tag{4.37}$$

This should be done as an exercise.

The reader is certainly aware that in nuclear and elementary-particle physics rest energy can be converted into kinetic energy, this being the source of energy for the stars and for nuclear energy devices. This convertibility of rest energy into kinetic energy is not a strict consequence of relativity theory, but is certainly compatible with it and is not compatible with Newtonian theory.

7. Photons and Relativistic Optics

We now wish to apply the formalism we have constructed to various problems which arise in physics and astronomy, and in this section we consider relativistic optics, and, in particular, the aberration of light and the Doppler effect.

One of the most fundamental conclusions of modern physics is that electromagnetic radiation consists of photons, and that for radiation of frequency ν each photon has energy $h\nu$, where $h = 6.625 \times 10^{-27}$ erg sec, and has momentum $P = h\nu/c$. These facts can be used to construct the four-momentum of a photon. Note that we cannot use the definition given by equation (4.31) since the photon travels at the speed of light ($ds = 0$) and its rest mass is zero.

However, since $m_o = 0$ for the photon, we can immediately use equation (4.37) with $P = h\nu/c$ to obtain

$$(P^a) = \frac{h\nu}{c}(n_x, n_y, n_z, 1), \tag{4.38}$$

where $\hat{n}$ is the unit three-vector in the direction of motion of the photon and ν is the observed frequency of the photon. Since $\hat{n}^2 = 1$ we have the relation

$$P^a P_a = 0. \tag{4.39}$$

This should be compared with equation (4.32).

Let us use the SLT equation (4.6) and substitute equation (4.38) into equation (4.13). For a=4, it follows that

$$P'^4 = \frac{h\nu'}{c} = \Lambda^4{}_1 P^1 + \Lambda^4{}_2 P^2 + \Lambda^4{}_3 P^3 + \Lambda^4{}_4 P^4 = \frac{h\nu}{c}\Gamma(1-\beta n_x),$$

or

$$\nu' = \nu\Gamma(1-\beta n_x) = \nu\Gamma(1-\vec{\beta}\cdot\hat{n}). \tag{4.40}$$

This is the famous Doppler effect formula. Note that if the photon is traveling in the same direction as 0', we have $\hat{n} = (1,0,0)$, and $\nu' = \nu[(1-\beta)/(1+\beta)]^{1/2} < \nu$, and 0' sees the photon as having a lower frequency than 0; i.e., the photon appears "red-shifted" in 0'. If the photon is traveling in the opposite direction from 0', then $\hat{n} = (-1,0,0)$, and

$$\nu' = \nu[(1+\beta)/(1-\beta)]^{1/2} > \nu \,, \tag{4.41}$$

and the photon appears "blue-shifted" in 0'.

We now refer back to our treatment of the QSO outbursts in Section (4.2), where we derived an equation for the observed time at the origin of 0 between the two outbursts. One can imagine a photon whose period, ν^{-1}, is equal to Δt^{obs} traveling along with the outburst signals from 0' to 0. Then $\nu' = (\Delta t')^{-1}$ and $\nu = (\Delta t^{obs})^{-1}$, and the expression for Δt^{obs} becomes equivalent to the "blue-shift" equation of the preceding paragraph. Note that $\Delta t'(\nu')$ and $\Delta t^{obs}(\nu)$ are observed by single clocks at rest in their respective SICS's.

To obtain the aberration formulas one sets a=1 in equation (4.13) to get $P'^1 = \frac{h\nu'}{c}n'_x = \Lambda^1{}_1 P^1 + \Lambda^1{}_4 P^4 = \frac{h\nu}{c}\Gamma(n_x - \beta)$, and then substitute equation (4.40) for ν' to obtain

$$n_x' = \frac{n_x - \beta}{1-\beta n_x} \,. \tag{4.42}$$

Similarly, finding P'^2 yields

$$n'_y = \frac{n_y}{\Gamma(1-\beta n_x)} \,. \tag{4.43}$$

A similar formula for n'_z may be found, and one may easily ascertain that $\hat{n}'^2 = 1$.

The aberration formulas are often presented in such a way that $n_z = 0$ and also so that the photon motion is toward the origin from the $x > 0$, $y > 0$ quadrant. Then $n_x = -\cos\theta$, $n_y = -\sin\theta$, and equations (4.42), and (4.43) become

$$\cos\theta' = \frac{\cos\theta + \beta}{1+\beta\cos\theta}, \tag{4.44}$$

$$\sin\theta' = \frac{\sin\theta}{\Gamma(1+\beta\cos\theta)}. \tag{4.45}$$

The Doppler effect formula has been verified up to terms of second order in β in the Ives-Stilwell experiment (see Schwartz, Section 3-3). However, the aberration formulas have been verified only to terms of first order in β, by astrometric methods.

The earth's velocity in orbit about the center of mass of the Solar System is $v \simeq 30$ km/sec, so that $|\beta| \simeq 10^{-4}$. Let the 0 system be at rest with respect to the c.m. of the Solar System, and let 0' be attached to the earth. Then for a given star, there are two times each year when we may set $\theta = \pi/2$. The "aberration angle" $\Delta\theta$ is then given by equation (4.44) and, assuming $|\Delta\theta| << 1$, we have $\cos(\pi/2 + \Delta\theta) = \cos\frac{\pi}{2}\cos\Delta\theta - \sin\frac{\pi}{2}\sin\Delta\theta \simeq -\Delta\theta \simeq +\beta \simeq 10^{-4}$. Thus $\Delta\theta \simeq -10^{-4}$ rad $\cong -20''$. This apparent change in position on the celestial sphere is easily measurable by modern astrometric methods.

8. Two Applications of Relativistic Optics: Observations of Relativistically Moving Objects; the Cosmological Doppler Effect, and Spectral Flux Intensities

In this section, we consider two astrophysically important applications of relativistic optics, the first being to observations of an object which is moving (or expanding) at a speed close to the speed of light.

From time-to-time in the astronomical literature, there have appeared assertions that an object had been observed moving faster than the speed of light, since a transverse velocity had been computed for it and shown to be > c. Such conclusions have been drawn with regard to wisps seen moving near the center of the Crab Nebula and with regard to alleged changes of position of certain radio sources. We will see that such conclusions are not necessarily justified.

Suppose we know the distance D to a small object which is moving at some unknown speed v with respect to us, and at some unknown angle μ to our line of sight (see Figure (4.2)) We, at point A, take two successive photographs of the object against the background of the "fixed" stars and note that the object has changed position by an angle $\Delta\theta$. We assume that $\Delta\theta$ is very small, and that the total distance traveled by the object is much less than D. Then if Δt is the time interval between the two photographs, we may compute an apparent transverse velocity $v_t \equiv D\Delta\theta/\Delta t$.

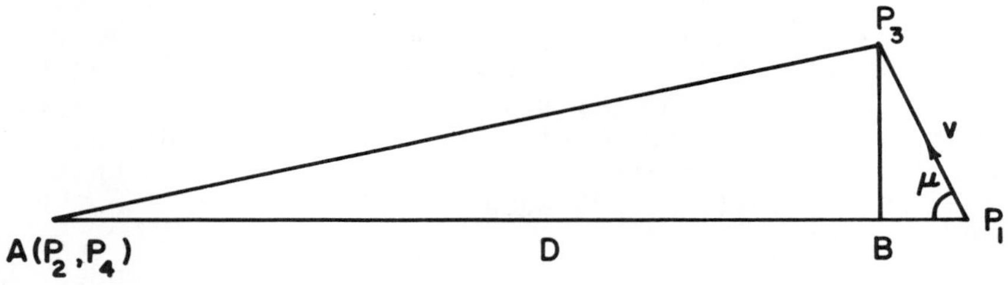

Fig. 4.2 The relativistically moving object. Photons are emitted by the object at events P_1 and P_3, and photographed at A as events P_2 and P_4. We assume $D \gg v\delta t$.

The question is this: Is it possible that $v_t > c$, even though $v < c$? The answer, surprisingly, is "yes"! Consider the following four events: Let P_1 be the event of emission of the photons which later arrive at A to produce the first photograph, the taking of which is event P_2, and let P_3 be the event of emission of the photons of the second photograph and P_4 be the taking of the second photograph. Then $\Delta t = t_4 - t_2$. If we abbreviate $\delta t \equiv t_3 - t_1$, the distance AP_1 is $AB + v\delta t\cos\mu$, and since $D \gg v\delta t$, we set $AP_3 \simeq D$, $AB \simeq D$. But since $t_2 = t_1 + AP_1/c$, and $t_4 = t_3 + AP_3/c$, we have $\Delta t = \delta t(1 - \beta\cos\mu)$. Further, $D\Delta\theta \simeq BP_3 = v\delta t\sin\mu$. Substituting these expressions, we obtain $v_t \simeq v\sin\mu(1-\beta\cos\mu)^{-1}$.

Note that it is definitely possible for v_t to exceed c. For example, for fixed $v < c$, the maximum v_t occurs for $\mu^{max} = \cos^{-1}\beta$, whereupon $v_t^{max} = v(1 - \beta^2)^{-1/2} = \gamma v$. Thus, as $v \to c$, $v_t^{max} \to \infty$, and $\mu^{max} \to 0$. This effect does not depend on the use of the LT, but is a consequence of the fact that the speed of light is a constant, finite quantity. It might not be obvious that we have not committed some profound error in using the approximation made above. However, the problem can be done exactly, and one can show that our result for v_t becomes exact in the limit $D \to \infty$ for fixed $v\delta t$.

Now consider an expanding luminous sphere, emitting photons continuously in all directions from each point on its surface. Since μ now covers all angles, the total extent of the sphere in the photograph is dictated by μ^{max}. Thus a relativistically expanding sphere appears to an observer to be expanding at a speed $v_t = v_t^{max}$. One can show that in this case $v_t > c$ if $v > c/\sqrt{2}$.

We now consider a problem which arises in cosmology, and which can be handled in part with the theory of special relativity; namely, the transformation properties of a spectral flux intensity $F(\nu)$, which gives the flux of energy in ergs cm^{-2} sec^{-1} Hz^{-1} received by an observer at frequency ν from a distant object. $F(\nu)d\nu$ is the energy flux received in the frequency interval $(\nu, \nu + d\nu)$. Often, $F(\nu)$ is expressed in "Janskys" (formerly called "flux units"), with 1 Jansky (1 flux unit) equal to 10^{-23} erg $cm^{-2}sec^{-1}Hz^{-1}$ and to 10^{-26}Watt $m^{-2}Hz^{-1}$.

Our problem is to apply the SLT to $F(\nu)$, under the assumption that the flux is completely collimated and is directed in the negative x direction. In astronomy, the motivation for this might arise in the following way: The spectral intensity function of a QSO (or cosmologically distant galaxy) is measured, and systems of identifiable spectral lines are seen, as peaks or valleys in $F(\nu)$. What we wish to do ultimately is to relate the observed Doppler-shifted $F(\nu)$ to the QSO's intrinsic spectral luminosity $L(\nu')$ in ergs sec^{-1} Hz^{-1}, as seen by an observer at rest at the QSO. Obviously, $L(\nu')$ can be related directly to the radiation processes operating in the QSO.

We note that the spectral lines are all red-shifted by the same amount $z \equiv (\lambda_{obs} - \lambda_o)/\lambda_o$, where λ_{obs} is an observed wavelength of a spectral line and λ_o is the "proper wavelength" of the line, as would be measured in the laboratory. Then if $\nu_o = c/\lambda_o$ is taken to be the frequency observed by 0', so that 0' and the QSO are at rest with respect to each other (cosmological transverse velocities are zero), the redshift z corresponds to a velocity such that $1 + z = [(1 + \beta)/(1 - \beta)]^{1/2}$ (set $\nu' = \nu_o$ and $\nu = \nu_{obs}$ in equation (4.41)), or

$$\beta = \frac{(1+z)^2 - 1}{(1+z)^2 + 1} . \tag{4.46}$$

Thus, an observer 0' moving past us with this velocity toward the QSO would see the spectral lines all blue-shifted back to their proper places in the spectrum.

The next question to answer is, what spectral intensity $F'(\nu')$ is observed by 0'? In considering this, it is convenient to work in terms of a spectral photon flux $n(\nu) = F(\nu)/(h\nu)$ photons cm^{-2} $sec^{-1}Hz^{-1}$, where h is Planck's constant. We consider a group of photons in the frequency interval $(\nu, \nu + d\nu)$ traveling together in an imaginary box,

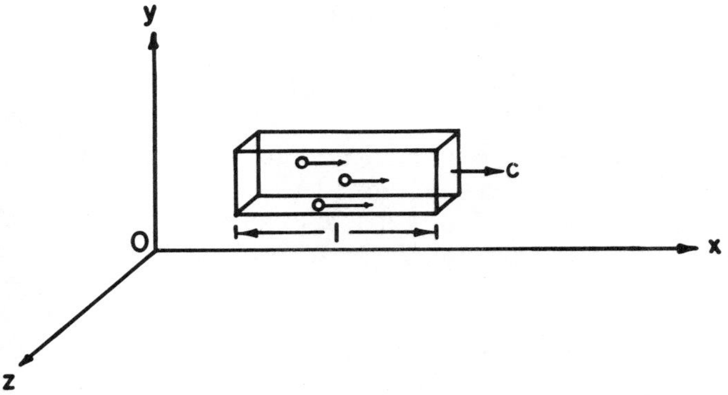

Fig. 4.3 Photons with frequencies between ν and $\nu+d\nu$ traveling in an imaginary box of length l and front surface area S.

whose length parallel to the direction of motion is ℓ and whose surface area perpendicular to the direction of motion is S (see Figure (4.3).

The number of photons in this group is then $n(\nu)\ell S d\nu/c$. This same group of photons is observed by O', who counts the same number $n'(\nu')\ell' S d\nu'/c = n(\nu)\ell S d\nu/c$, where ν' given by equation (4.41). The surface area S is, of course, measured to be the same by both observers. Therefore, $n'(\nu')\ell' d\nu' = n(\nu)\ell d\nu$. We cannot apply the Lorentz contraction formula to ℓ and ℓ', since the box is not at rest in either O or O'. In fact, since the box is moving with the photons (it could itself be constructed out of special "marker photons"), there exists no SICS with respect to which it is at rest.

Now let P_1 be the event of the O' origin meeting the front of the box, and P_2 be the event of the O' origin meeting the rear of the box. Since $x_2' = x_1'$, $\Delta t' = t_2' - t_1' = \ell'/c$ is a proper time interval, and eqn. (4.4) yields $\Delta t = \Gamma\ell'/c$. We must now express Δt in terms of ℓ. The velocity of the box with respect to O' as measured in the O system is c + V, and therefore $\Delta t = \ell/(c + V)$. (The reader is invited to ponder this last sentence, which could be set up as an elementary algebra problem.) The conclusion is then that $\ell = \ell'\Gamma(1 + \beta) = \ell'[(1 + \beta)/(1 - \beta)]^{1/2} = \ell'(1+z)$. Equation (4.41) then implies that $d\nu' = \ell d\nu/\ell'$, so that we finally obtain $n'(\nu') = n(\nu)$. Using equation (4.41) again as $\nu' = \nu(1+z)$, we write $F'(\nu') = h\nu' n'(\nu') = h(1+z)\nu n(\nu)$, or

$$F'(\nu') = (1 + z)F(\nu) . \tag{4.47}$$

If we know how far away the QSO was when it emitted the photons which we received at 0, we can find the total spectral luminosity $L'(\nu')$ in erg $sec^{-1}Hz^{-1}$measured in the QSO's rest frame. To do this, we must find the distance d' from the QSO to the point of reception of the photons as seen from 0'. Then, assuming that the QSO radiates isotropically in its own rest frame 0', we have $L'(\nu') = 4\pi(d')^2F'(\nu')$. To find d' in terms of d, the distance to the QSO seen from 0 at the time of emission, is a simple exercise in the use of the SLT. Let P_1 be the event of emission of the photons, and let P_2 be the event of their reception at 0. It is convenient to assume that the origins of 0 and 0' are coincident at the occurrence of P_2, at t=0. Then $x_1 = d$, $t_1 = -d/c$, and $d' = x_1'$. Equations (4.1) then imply $d' = \Gamma(x_1 - Vt_1) = \Gamma(1 + \beta)d = (1 + z)d$. Using equation (4.47) yields

$$L'(\nu') = 4\pi d^2(1 + z)^3 F[\nu'/(1 + z)]. \tag{4.48}$$

Note that we are again not able to use the Lorentz contraction formula because P_1 and P_2 are not simultaneous events with respect to either frame.

In the cosmological literature, a distance d_L is defined such that $L \equiv 4\pi d_L{}^2F$ where $L \equiv \int L'(\nu')d\nu'$, and $F \equiv \int F(\nu)d\nu$. This d_L is called the "luminosity distance". See, for example, the book by Weinberg (pg. 421). Multiplication of equation (4.48) by $d\nu'$ and integrating, with a change of variable on the right-hand side, yields $L = 4\pi d^2 (1 + z)^4F$, so that our d and the "canonical" d_L are related by $d_L = d(1 + z)^2$. Also, we have $d_L = d'(1 + z)$. It is difficult to see any rationale for the use of d_L , except as convenient parameter. However, an accurate discussion of cosmology requires the use of a gravitation theory, such as the general theory of relativity, and is beyond the scope of this book.

5. Synchrotron Spectra

5.1 Introduction

In this chapter we will be interested in processes leading to emission of radiation or affecting propagation of radiation in the range of frequencies 10^7 Hz - 10^{11} Hz - this is more or less the extent of the spectrum within which the earth's atmosphere is transparent to radio radiation. The high frequency edge of this spectral region (radio window) is due mainly to infrared absorption bands of water and carbon dioxide. The low frequency side is determined by the critical frequency of reflection by the ionosphere. This frequency depends on the ionospheric electron density, which is variable (with the time of day, geographic location of the observer, solar activity, etc.), and the observations can occasionally be made down to 10^6 Hz.

Our main emphasis will be on the radiation produced by relativistic electrons gyrating in a magnetic field (synchrotron radiation). We will investigate some properties of its distinctive spectrum, as the synchrotron mechanism is thought at the present time to have important applications to cosmic radio sources.

All units are CGS unless otherwise specified.

5.2 Mechanism of Emission of Electromagnetic Radiation

A charge moving uniformly in a straight line in vacuum cannot absorb or emit radiation because the laws of conservation of energy and momentum cannot be simultaneously satisfied in such interaction with radiation. Consider e.g.

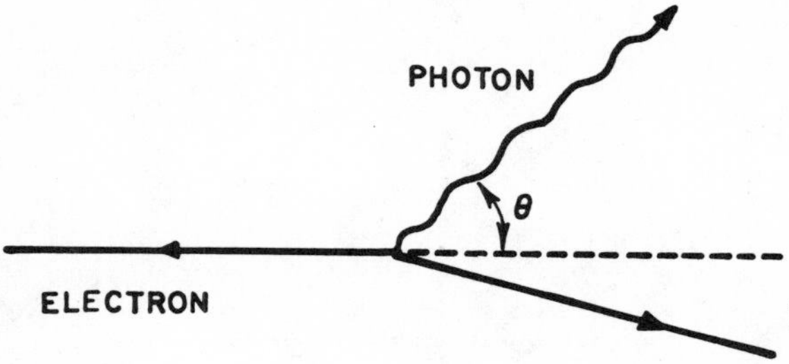

Fig. 5.1 Emission of a photon by an electron

the emission of a photon (velocity $u = c$, momentum $|\vec{p}_{ph}| = h\nu/c$, energy $h\nu$) by an electron (velocity v, momentum p, energy E, before emission; p', E', after emission), as represented in Figure 5.1. The conservation of energy and momentum laws are:

$$E - h\nu = \sqrt{(p')^2c^2 + m^2c^4}\,, \qquad \vec{p} - \vec{p}_{ph} = \vec{p}'\,.$$

Squaring both the above equations and subsequently eliminating $(p')^2$ gives

$$2c^2pp_{ph}\cos\theta = c^2p_{ph}^2 - (h\nu)^2 + 2h\nu E$$

(since $E^2 = p^2c^2 + m^2c^4$), and since $p_{ph} = h\nu/c$ and $pc^2 = vE$ we have

$$\cos\theta = \frac{c}{v} > 1, \tag{5.1}$$

which means that no photon can be emitted by a uniformly moving charge in vacuum. Only scattering of radiation by electrons (moving uniformly) in vacuum is possible; this scattering can be considered as an absorption process followed immediately by an emission process.

In a medium (even homogeneous), however, capable of propagating radiation, rectilinear and uniform motion of a charge will produce radiation under certain conditions. Indeed, if the phase velocity of propagation of the radiation in this medium is $u < c$ (this is, for example, the case of the extraordinary mode in a magneto-active plasma, see Section 5.7), the requirement of conservation of energy and momentum can be satisfied and the expression (5.1) takes the form

$$\cos \theta = \frac{u}{v} < 1, \qquad (5.2)$$

for appropriate frequency range and velocity of the charge. In such a situation radiation will be emitted at an angle θ; this radiation called Čerenkov (or Vavilov - Čerenkov) radiation is analogous to a ship's bow wave or a Mach wave at supersonic velocities. Čerenkov radiation is generated at the expense of kinetic energy of the charge on which a retarding force acts as a recoil resulting from radiation of momentum. It can be looked upon as emission produced through a cooperative phenomenon involving a large number of particles in the medium which are accelerated by the fields of the moving charge and therefore emit radiation. There is always an upper limit to the frequency of Čerenkov radiation for a given charge velocity v, since the propagation velocity in any medium u approaches that in vacuum when the frequency is high enough.

In a vacuum it is therefore necessary to accelerate a charge (to change the direction of velocity or its absolute value, or both) in order to produce emission of radiation. This is easy to see if one visualizes radiation as an oscillation of electromagnetic field, which is connected to matter through charges - an oscillation of a charge due to a force will therefore entail an oscillation of the field. Acceleration of charges may be due to a number of forces. If the forces acting on a charge are strong or weak interactions, one has an emission of radiation ususally in the γ-ray range called radioactivity. Gravitational forces do not produce oscillations of charges at radio frequencies. Such oscillations can, however, be caused by electromagnetic forces.

Analogous to radiation of a non-uniformly moving (accelerated) charge in vacuum is radiation emitted by a charge moving uniformly along a straight line in an electrically inhomogeneous medium (i.e. with a variable dielectric constant), called transition radiation. In both cases radiation is due to the change in relationship between the phase velocity of electromagnetic waves and the velocity of the charge. In non-uniform motion in vacuum the charge velocity changes, in uniform motion in an inhomogeneous medium the phase velocity of the wave changes. As the result the field connected with the charge becomes detached from it and the radiation takes place. In addition to the radiation connected with the variations of the Coulomb field of an electron as it moves through a medium the properties of which vary from point to point, the term transition radiation includes also the scattering on fluctuations of the dielectric constant of Čerenkov waves emitted by the charge. The first process, unlike the Čerenkov process, is not necessarily a relativistic effect and can be observed at low velocities of the moving charge, while Čerenkov radiation takes place only when the charge velocity is larger than the phase velocity of the wave. The first effect is important in the frequency range where there is no Čerenkov effect. Historically the term transition radiation was used first to describe a particular case of radiation emitted by an electron moving uniformly through an interface between two media of different electrical properties. This case may have astrophysical applications as some emission of soft X-rays and ultraviolet radiation in interstellar space, supernova remnants and some other objects might be due to cosmic ray electrons passing through dust grains.

Let us now return to the radiation of charges oscillating in vacuum, neglecting the mutual effect of neighboring charges or neighboring groups of charges. If N charges are oscillating independently the radiation is incoherent. The intensity of radiation of such an ensemble of charges is the sum of intensities i of radiation of individual particles

$$I = Ni . \tag{5.3}$$

If there is a spatial non-uniformity of the distribution of current in the source, e.g. consisting of grouping of N_c changes within dimension l much smaller than the wavelength λ of radiation (both lengths l and λ being measured in the same laboratory system of reference), the charges radiate in phase in all directions; such radiation is called <u>coherent radiation.</u> Requirement $l \ll \lambda$ assured that the difference in phases between radiating particles in the group is small. The intensity of radiation of such a group of charges is

$$I_g = N_c^2 i, \tag{5.4}$$

since the intensity of the radiation of a single charge is proportional to the square of the charge $i \propto e^2$ (since $i \propto E^2$, and the electric field $E \propto e$); and a group of N_c charges is here radiating as a charge N_c e. If we have N_g such groups of N_c charges (a total of N charges, $N = N_g N_c$) and the groups are radiating not in phase, the intensity of resulting radiation is

$$I = N_g N_c^2 i = N_c \cdot Ni > Ni \quad , \tag{5.5}$$

if $N_c > 1$. The intensity of the radiation from such a system of charges is larger than the intensity of radiation of N incoherently radiating charges by a factor N_c which is a number of charges radiating coherently in a group.

We define a coherent mechanism of radiation as mechanism which yields the radiation of intensity

$$I > Ni \tag{5.6}$$

as opposed to an incoherent mechanism, for which $I = Ni$. It should be noted that a coherent mechanism may yield in general incoherent or partially coherent radiation. The grouping of radiating particles, described above, is an example of a coherent antenna-type mechanism of radiation; the name derives from the fact that the radiation of an antenna is coherent in the direction perpendicular to the axis of a thin wire or to the surface of a thin disc of an antenna.

In astrophysical conditions the antenna-type coherent mechanism of radiation encounters difficulties due to the stability of coherently radiating groups of charged particles. Even a small velocity distribution of charges within a group will quickly lead to a diffusion of the group and to the failure of the coherency condition $l < \lambda$ at meter and shorter wavelengths.

In the preceding discussion we have neglected the mutual effect of neighboring charges or neighboring groups of charges which can lead to reabsorption of radiation, and therefore to decrease of intensity as compared to that given by equations (5.3) or (5.6). In certain cases, however, the interaction with neighboring radiators may produce an effect of negative reabsorption, i.e. of amplification of radiation stimulated by the inversion in populations of energy levels. The intensity of radiation produced as a result of this mechanism, called coherent maser-type mechanism of radiation, is

$$I = I_o\, e^{-\kappa s} , \tag{5.7}$$

where κ, the absorption coefficient, is negative in the range of this mechanism, producing amplification of radiation, s is path-length in a uniform medium and I_o is the incident intensity.

We will discuss now the radiation of a single charge accelerated by a varying electric field $\mathcal{E}$. The charge e of mass m radiating as a dipole emits power

$$P = \frac{2}{3c^3}\ddot{d}^2 ;$$

the motion of the charge due to the varying electric field,

$$\ddot{x} = \frac{e}{m}\mathcal{E} ,$$

produces variations in the dipole moment d = ex. The mean power radiated is therefore

$$P = 2\left(\frac{8\pi}{3}\frac{e^4}{m^2c^4}\right) \cdot c \cdot \left\langle \frac{\mathcal{E}^2}{8\pi} \right\rangle = 2\sigma c\langle u_{\mathcal{E}}\rangle , \tag{5.8}$$

σ and $u_{\mathcal{E}}$ are the Thomson scattering cross-section and the energy density of the electric field $\mathcal{E}$. If the electron moves with relativistic velocities, the energy density of the electric field $u_{\mathcal{E}}$ entering equation (5.8) should be expressed in the electron rest frame, and

$$P = 2\sigma c\gamma^2 \frac{(\vec{\mathcal{E}}+\vec{\beta}\times\vec{H})^2}{8\pi} , \tag{5.9}$$

where $\beta = v/c$ and $\gamma = (1-\beta^2)^{-\frac{1}{2}}$; P is the energy received by the observer in Δt of the observer's time.

Let us consider now an electron under the influence of the electromagnetic field of a beam of photons with energy density u_{rad}. Let us assume that the electron is non-relativistic ($\beta \simeq 0$, $\gamma \simeq 1$); this is the case of Thomson scattering of radiation on electrons. The scattered power is

$$P = \sigma c u_{rad} ; \tag{5.10}$$

since half of the energy density of radiation is in the form of energy associated with the magnetic field, $\mathcal{E}^2/8\pi = 1/2 u_{rad}$.

If the electron is relativistic, the low energy photon will acquire energy from the high energy electron during scattering. This process is called inverse Compton scattering since the energy is transferred from electrons to photons, contrary to the classical Compton (Thomson) case in which the energy transfer takes place in the opposite direction. If the distribution of photons is isotropic, the mean power scattered per electron is

$$P \cong \gamma^2 \sigma c u_{rad}. \tag{5.11}$$

The average energy (frequency) of a scattered photon is γ^2 times its original energy (frequency).

A low energy charge ($v \ll c$) gyrating in a magnetic field on a circular orbit of radius (radius of gyration)

$$r_G = \frac{v_\perp}{\omega_G} , \tag{5.12}$$

with the frequency (gyrofrequency, Larmor frequency)

$$\omega_G = \frac{eH}{mc} = 1.8 \times 10^7 H, \tag{5.13}$$

emits radiation called cyclotron radiation or gyroradiation. At large distances this radiation is of dipole character. If there is no motion along the field, as we assume for simplicity, the radiation of the electron is equivalent to the radiation of two mutually perpendicular linear oscillators shifted in phase by $\pi/2$. The frequency of the radiation is the same as the frequency of gyration, that is ω_G. The intensity, averaged over a period, is fairly isotropic; the polarization is linear in the plane of the orbit, circular perpendicular to the orbit, and elliptical in any other direction. If the motion along the field is non-relativistic, the intensity is nearly the same as in the case of a circular orbit. The total power radiated is given by equation (5.9) ($\gamma \simeq 1$):

$$P = 2\sigma c \frac{(\bar{v}\times\bar{H})^2}{8\pi c^2} = \frac{4}{3}\,\frac{e^4 H^2 \sin^2\theta}{m^2 c^5}\,E = \frac{2}{3}\,\frac{e^2}{c^3}\,\omega_G^4\;r_G^2\,, \qquad (5.14)$$

where θ is the pitch angle, that is the angle between v and H, and E is the energy of the charge.

If the energy of a charge gyrating in a magnetic field is relativistic ($\gamma \gg 1$), the radiation emitted, called synchrotron radiation, has much higher power:

$$P = 2\sigma c\gamma^2\beta^2\,\frac{H_\perp^2}{8\pi} \cong 2\sigma c\gamma^2 u_H \sin^2\theta = 2.4\times 10^{-3} H_\perp^2 E^2\,, \qquad (5.15)$$

where E is the energy of the charge , u_H is the magnetic energy density, and θ is the pitch angle (the angle between v and H). β is well approximated by unity. The properties of synchrotron radiation, which are very different than those of cyclotron radiation, will be discussed in more detail in the next section.

Thomson scattering and inverse Compton scattering are the forms of radiation of a (nonrelativistic and relativistic, respectively) charge in a field of an electromagnetic wave of low intensity, that is when

$$f = \frac{\omega_G}{\nu_W} \ll 1\;; \qquad (5.16)$$

ω_G is the (non-relativistic) gyro frequency of a charge e of mass m in the field of the wave H, $\omega_G = eH/mc$, and ν_W is the frequency of the wave. As f becomes large, the character of the emission changes substantially. The charge moving with velocity v cos α relative to the direction of (plane) wave velocity u (we will assume $\alpha \neq 0$) encounters wave crests every Δt in the laboratory frame, (see Figure 5.2)

$$\Delta t = \frac{\lambda_w}{|u-v\cos\alpha|} = \frac{1}{\nu_w(1-\beta\cos\alpha)} \simeq \frac{1}{\nu_w}, \tag{5.17}$$

λ_w is the wavelength. During this time Δt the trajectory of the charge is deflected by an angle $\psi = \omega_G \Delta t/\gamma = \omega_G/\gamma\nu_w$. The beam width of the radiation emitted by a relativistic charge is $\psi \sim 1/\gamma$; therefore, if $f > 1$, the beam of the radiation sweeps past the observer in a way analogous to a charge emitting synchrotron radiation, i.e. the duration of the pulse is short compared to pulse separation, see Section 5.3. The radiation of a charge in the field of a strong electromagnetic wave will then have properties closer to synchrotron radiation than to inverse Compton radiation. This radiation, which can be important in the vicinity of pulsars, is referred to as synchro-Compton radiation or non-linear inverse Compton radiation. The angular deflection ψ in Δt corresponds to an average circular frequency $\omega_{sc} \sim \psi/\Delta t = \omega_G/\gamma$ which is independent of the wave frequency; and the frequency at which the bulk of radiation is received (critical frequency) is $\omega_G\gamma^2$, as in the case of synchrotron radiation. In inverse Compton scattering this critical frequency is $\omega_w\gamma^2$, the ratio of these two critical frequencies is therefore of the order of f. We referred here to the charge as having relativistic velocity; indeed, when $f \gg 1$ even an initially nonrelativistic charge will become relativistic in any chosen frame at some phase of its periodic motion, and will act in transforming the energy of intense very low frequency waves into higher frequency radiation. Synchro-Compton radiation differs substantially from synchrotron radiation in its polarization. Circular polarization of synchro-Compton radiation is of the order of $1/f$ when the low frequency wave is polarized circularly.

There are other processes of emission of radiation, corresponding to other ways of accelerating charges. If charges are acted upon by electrostatic fields we will have bremsstrahlung and emission of plasma waves. Charges can also be oscillating in the field of hydromagnetic waves, and so on. Among the mechanisms mentioned bremsstrahlung (free-free or thermal radiation) is of great importance to astrophysics and was discussed in detail in the first two chapters.

5.3 Synchrotron Radiation

In a reference system in which the electron velocity is relativistic, $v_\perp \sim c$ (we are assuming for a time being a circular orbit), $\beta \cong 1$, the electron gyrates with the frequency

$$\omega_H = \frac{\omega_G}{\gamma} \quad \text{(much smaller than } \omega_G\text{)}, \tag{5.18}$$

and has a radius of gyration

$$r_H = \frac{v_\perp}{\omega_H} \cong \frac{c}{\omega_H} = \frac{c}{\omega_G}\gamma \qquad \text{(much larger than } r_G\text{)}. \tag{5.19}$$

If we transform the frequency and the intensity of the cyclotron radiation into the observer's frame, in which the electron velocity is relativistic, we have (see Section 4.7)

$$\omega = \frac{\omega_G}{\gamma(1-\beta\cos\psi)}, \qquad I = \frac{I_{cyc}}{\gamma^3(1-\beta\cos\psi)^3},$$

where ψ is the angle between the velocity and the direction of wave propagation.

If $\psi \lesssim \frac{1}{\gamma}$, $\quad \omega \sim \frac{\omega_G}{\gamma}\gamma^2 = \omega_H\gamma^2, \quad I \sim I_{cyc}\gamma^3$,

and if $\psi \gg \frac{1}{\gamma}$ $\quad \omega \sim \frac{\omega_G}{\gamma} = \omega_H \quad I \sim I_{cyc}/\gamma^3$;

since $1-\beta\cos\psi \simeq 1-\beta + \beta\frac{\psi^2}{2}$ is of the order of $1-\beta = 1/\gamma^2$ if $\beta\frac{\psi^2}{2} \simeq 1-\beta$, that is if $\psi \simeq \sqrt{1-\beta} = 1/\gamma$. In the ultrarelativistic case ($\gamma \gg 1$) the radiation is therefore very directional: both frequency and intensity of radiation are very much higher within a very small cone around v, up to angles of the order of $1/\gamma$; the observer sees short pulses when the electron moves directly at him.

The duration of this pulse can be determined by estimating the time, τ_e, during which the velocity vector remains within an angle $1/\gamma$ from the direction of the observer; from Figure 5.3 we have (in the observer's system)

$$\tau_e = \frac{1}{2\pi\gamma \cdot \frac{\omega_H}{2\pi}} = \frac{1}{\omega_G} , \tag{5.20}$$

or, what amounts to the same thing,

$$\tau_e = \frac{1}{v} = \frac{r_H}{v\cdot\gamma} = \frac{1}{\gamma\omega_H} = \frac{1}{\omega_G} .$$

The pulse emitted within the time τ_e will be received by a distant observer during the time τ due to that part of the Doppler effect arising from the component of the electron velocity along the line of sight,

$$\tau = \tau_e(1-\beta \cos \psi) \sim \frac{\tau_e}{\gamma^2} = \frac{1}{\omega_G\gamma^2} \simeq \frac{1}{\omega_c} , \tag{5.21}$$

where

$$\omega_c = \frac{3}{2}\,\omega_G \sin \theta\gamma^2 = 4 \times 10^{19} H_{\perp} E^2 = 2\pi c_1 H_{\perp} E^2 , \tag{5.22}$$

(much larger than ω_G)

is called the critical frequency (θ is the angle between the velocity and the field--pitch angle). The interval between the two consecutive pulses is

$$T = \frac{2\pi}{\omega_H} = \frac{2\pi\gamma}{\omega_G} , \tag{5.23}$$

very large compared with the duration of a pulse.

All preceding formulae referred to a circular motion of the electron. In general, this motion will be helical if the parallel component of velocity does not vanish. In this

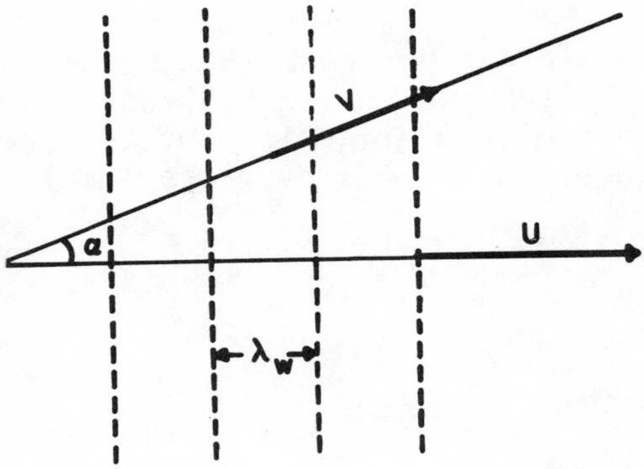

Fig. 5.2 Deriving equation (5.17)

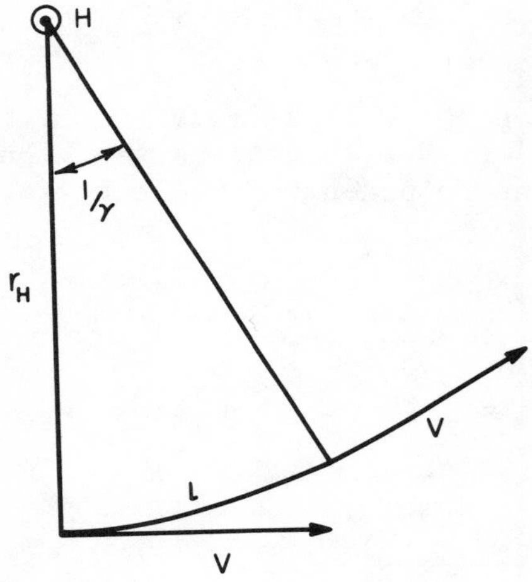

Fig. 5.3 Deriving equation (5.20)

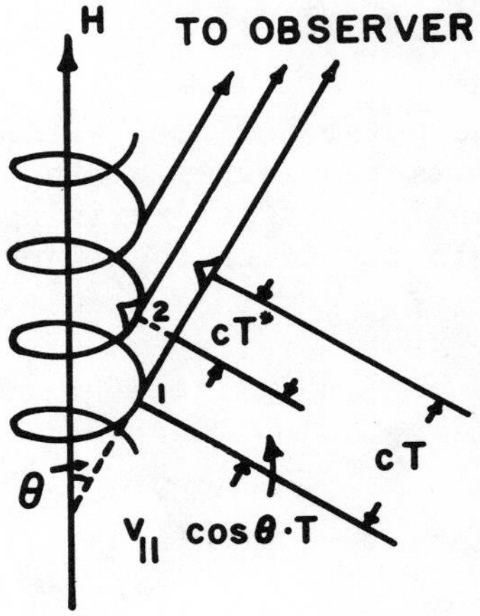

Fig. 5.4 Deriving equation (5.26)

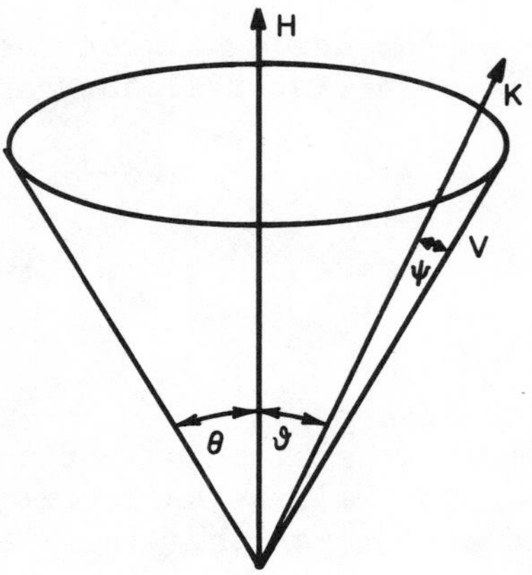

Fig. 5.5 Relationship between the direction of the magnetic field H, velocity v, and the direction toward the observer, k

general case in the equation for τ we have to replace r_H

$$r_H = \frac{v \sin\theta}{\omega_H} \cong \frac{E \sin^2\theta}{eH_\perp}, \tag{5.24}$$

the radius of gyration, by r_H^*,

$$r_H^* = \frac{v}{\omega_H \sin\theta} = \frac{E}{eH_\perp}, \tag{5.25}$$

the radius of spatial curvature of the trajectory. The expression for τ will be, after substituting r_H^* for r_H, the same as previously with H being replaced by $H_\perp$, i.e. by $H\sin\theta$. The expression for T will also have to be modified. If the direction of wave propagation makes an angle ϑ with the direction of the magnetic field, the pulses will follow each other not after a time T, but after a time T* (see Figure 5.4) such that

$$cT^* = cT - v_\parallel T \cos\vartheta \simeq cT - vT\cos^2\vartheta, \tag{5.26}$$

since the interval of time between the electron being at 1 and at 2 is equal to the period $T = 2\pi/\omega_H$ and the distance between 1 and 2 is $v_\parallel T = vT\cos\theta$. θ is the angle between v and H; $\theta \simeq \vartheta$ to have appreciable radiation, see Figure 5.5. Finally,

$$T^* = T(1 - \beta\cos^2\vartheta) \simeq T\sin^2\vartheta. \tag{5.27}$$

The electron motion must be relativistic not only in the observer's frame, but also in the frame commoving with the gyrocenter. Figure 5.6 (middle bottom) shows the electric field, associated with the synchrotron radiation, at a large

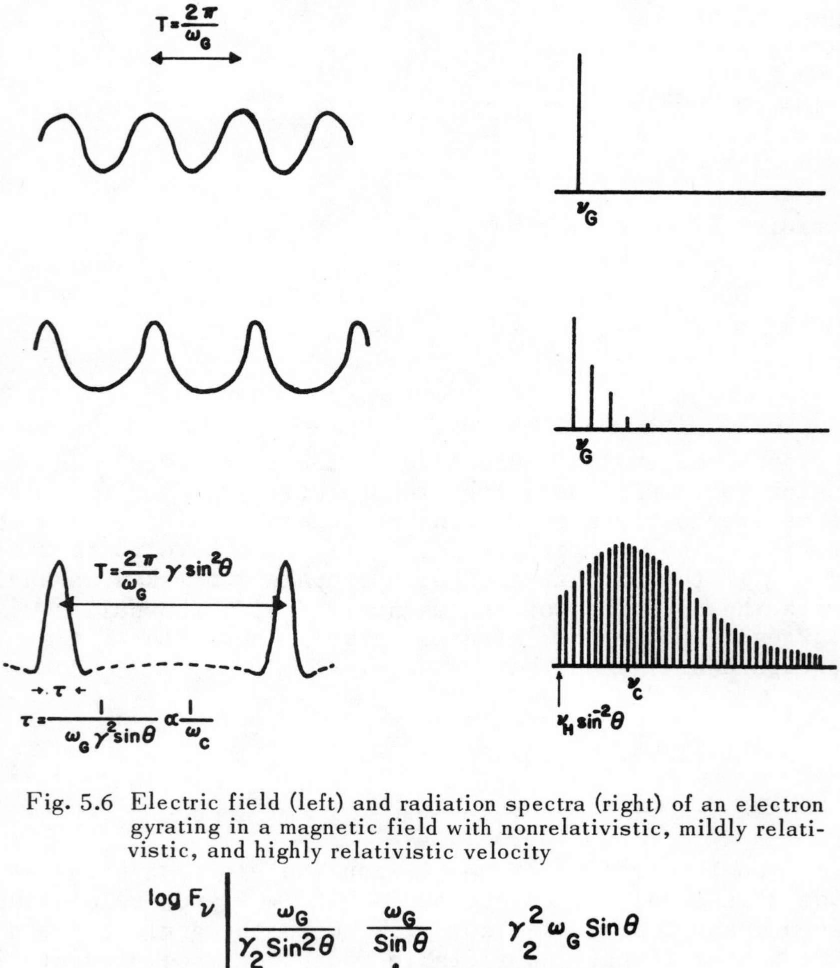

Fig. 5.6 Electric field (left) and radiation spectra (right) of an electron gyrating in a magnetic field with nonrelativistic, mildly relativistic, and highly relativistic velocity

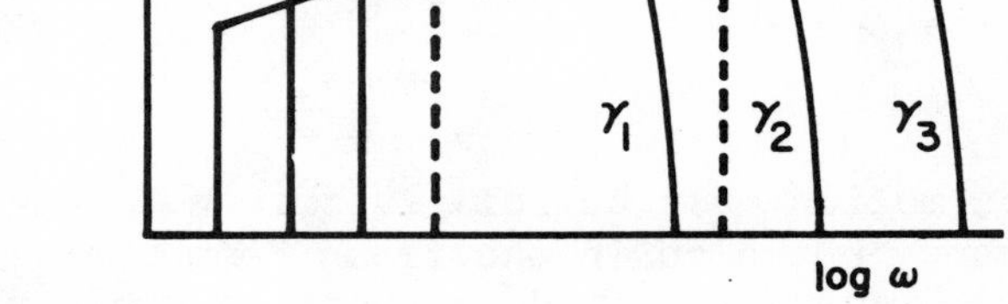

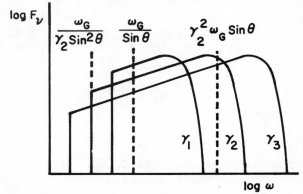

Fig. 5.7 Synchrotron spectra of several electrons with different energies

distance from the radiating electron. The duration of weak negative pulse is an order of magnitude longer than the duration of the positive peak. The mean electric field is, of course, zero. The frequency spectrum of the radiation can be obtained by a Fourier transform of the pulse train. It will consist of a series of harmonics of the fundamental frequency ω_F,

$$\omega_F = \frac{2\pi}{T^*} = \frac{\omega_H}{\sin^2\theta} , \qquad (5.28)$$

contained within an envelope determined by the individual pulse. In reality , many-electron interactions cause broadening of individual harmonics, and the spectrum of a single electron is practically continuous. The envelope will have a maximum at a frequency of the order of $1/\tau = \omega_c$, in fact at a frequency

$$\omega_{max} = 0.3\omega_c . \qquad (5.29)$$

The spectrum is shown in Figure 5.7. It starts at the fundamental frequency $\omega_F = \omega_G/\gamma \sin^2\theta$, rises as $\omega^{1/3}$ to the 0.3 times the critical frequency $\omega_c = \frac{3}{2}\omega_G\gamma^2\sin\theta = \frac{3}{2}\omega_F(\gamma\sin\theta)^3$, and falls off exponentially above that frequency.

Figure 5.6 shows that radiation spectra of an electron gyrating in a magnetic field with nonrelativistic, mildly relativistic, and highly relativistic velocity.

The polarization of the synchrotron radiation of an individual electron is elliptical, the major axis of the polarization ellipse is perpendicular to the projection of the field (Figure 5.8). The ellipticity depends on the angle ψ; the tangent of the ratio of the minor to the major axis is proportional to ψ. Polarization is linear in the direction $\psi = 0$. If the radiation comes from a distribution of electrons which is isotropic over the angle $\psi \sim 1/\gamma$, the circular contributions to polarization cancel out to the order of $1/\gamma$. As a function of frequency the degree of linear polarization varies from 0.5 at low to 1.0 at high frequencies; around ν_c, where most of radiation is emitted, the polarization is close to 0.7.

The total power emitted at all frequencies is (equation 5.15)

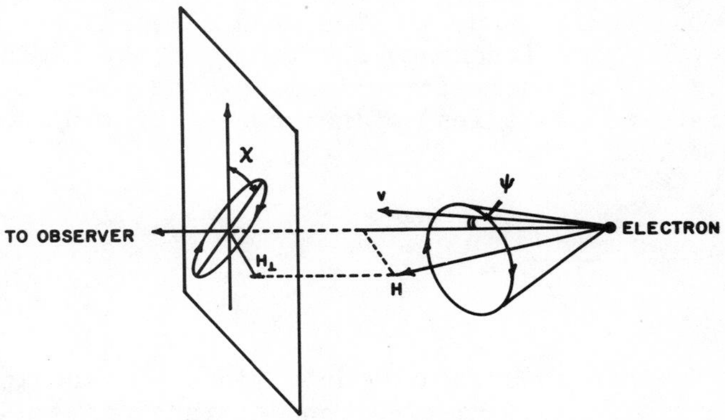

Fig. 5.8 Orientation of the polarization ellipse of synchrotron radiation of an electron

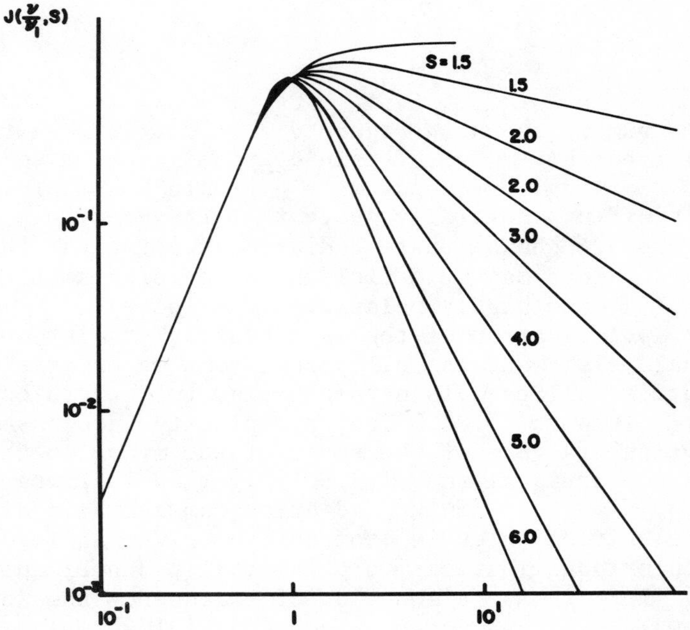

Fig. 5.9 The synchrotron spectrum of electrons with a power-law energy distribution (the function J)

$$P = \frac{2e^4 H_\perp^2}{3m^2c^3} \gamma^2 = 2.4 \times 10^{-3} H_\perp^2 E^2 = c_2 H_\perp^2 E^2 \ . \qquad (5.30)$$

For particles of mass M and charge Ze, the losses are $(M/Zm)^4$ smaller for a given energy. Synchrotron radiation of protons is therefore negligible compared with the radiation of electrons and positrons of the same energy E.

In radio astronomy we have very often to consider synchrotron radiation emitted by an ensemble of isotropic electrons with a power law distribtuion over their energies; the number of electrons N(E)dE with energies between E and E + dE being

$$N(E)dE = N_o E^{-s} \ dE, \qquad (5.31)$$

The emission coefficient for the synchrotron radiation of such electrons is

$$\varepsilon_\nu = \int N_o E^{-s} \ P_\nu(E) \ dE, \qquad (5.32)$$

where $p_\nu(E)$ is the emission of a single electron of energy E (cf. Figure 5.6). Since this emission p_ν decreases exponentially at $\nu > \nu_c$ and since in a power law distribution given by equation (5.33) the number of electrons with energies higher than the energy corresponding to ν_c decreases rapdily with increasing energy, one can assume in the following order-of-magnitude approach that each electron emits power P (equation 5.30) at the critical frequency ν_c (equation 5.22). We then have

$$\varepsilon_\nu \propto N_o H_\perp^2 \int E^{-s} E^2 \delta(\nu-\nu_c)\, dE$$

$$\propto N_o H_\perp \int E^{-s+1} \delta(\nu-\nu_c)\, d\nu_c ,$$

since $\nu_c \propto H_\perp E^2$, $d\nu_c \propto H_\perp E\ dE$. Therefore

$$\varepsilon_\nu = c_5(s)\, N_o H_\perp^{\frac{s+1}{2}} \left(\frac{\nu}{2c_1}\right)^{-\frac{s-1}{2}} . \tag{5.33}$$

More careful computations give the values of c_5 (s) presented in Table 5.1. The absorption coefficient for the synchrotron radiation of a power-law distribution of electrons is

$$\kappa_\nu = c_6(s) N_o H_\perp^{\frac{s+2}{2}} \left(\frac{\nu}{2c_1}\right)^{-\frac{s+4}{2}} . \tag{5.34}$$

The values of the constant c_6 (s) are given in Table 5.1.

To obtain the form of the spectrum one should solve the transfer equation

$$\frac{dI}{ds} = -\kappa_\nu I_\nu + \varepsilon_\nu . \tag{5.35}$$

The integral form of this equation is

$$I_\nu(S) = I_\nu(0) e^{-\tau_\nu(s,0)} + \int_o^s e^{-\tau_\nu(s,s')} k_\nu S_\nu ds' ; \; s_\nu = \frac{\varepsilon_\nu}{k_\nu} . \tag{5.36}$$

The solution for I_ν (0) = 0 and for s_ν independent of position within the source is

$$I_\nu(\tau_\nu) = S_\nu \int_0^s e^{-\tau_\nu(s,s')} d\tau' = S_\nu(e^{-\tau_\nu(s,s')} - e^{-\tau_\nu(s,0)}$$

$$= S_\nu(1-e^{-\tau_\nu}), \quad (5.37)$$

where $d\tau_\nu = k_\nu ds$. Substituting the expressions for the absorption and emission coefficients we have

$$I_\nu = S(\nu_1)\cdot J(\frac{\nu}{\nu_1}, s), \quad (5.38)$$

where ν_1, defined by the condition $\tau(\nu_1) = 1$, is given by (s is the extent of the radiation region)

$$\nu_1 = 2c_1(sc_6)^{\frac{2}{s+4}} N_o^{\frac{2}{s+4}} H_\perp^{\frac{s+2}{s+4}}, \quad (5.39)$$

$$S(\nu_1) = \frac{c_5}{c_6} H_\perp^{-1/2} \left(\frac{\nu_1}{2c_1}\right)^{5/2}. \quad (5.40)$$

The function J is given in Figure 5.9. Differentiating the equation (5.41) for the intensity and setting the result to be equal zero we can find the frequency at which the intensity attains a maximum. The equation for this frequency, ν_m is

$$e^{\tau_m} = 1 + \frac{\gamma+4}{5}\tau_m, \quad (5.41)$$

s	2	3	4	5
τ_m	0.35	0.65	0.88	1.08 .

TABLE 5.1

The Functions $c_5(s)$, $c_6(s)$ and $c_{14}(s)$

s	c_5	c_6	c_{14}
0.5	2.66 E-22	1.62 E-40	3.38 E 29
1.0	4.88 E-23	1.18 E-40	1.33 E 30
1.5	2.26 E-23	9.69 E-41	2.38 E 30
2.0	1.37 E-23	8.61 E-41	3.48 E 30
2.5	9.68 E-24	8.10 E-41	4.65 E 30
3.0	7.52 E-24	7.97 E-41	5.89 E 30
3.5	6.29 E-24	8.16 E-41	7.22 E 30
4.0	5.56 E-24	8.55 E-41	8.51 E 30
4.5	5.16 E-24	9.24 E-41	9.94 E 30
5.0	4.98 E-24	1.03 E-40	1.14 E 31
5.5	4.97 E-24	1.16 E-40	1.30 E 31
6.0	5.11 E-24	1.24 E-40	1.45 E 31

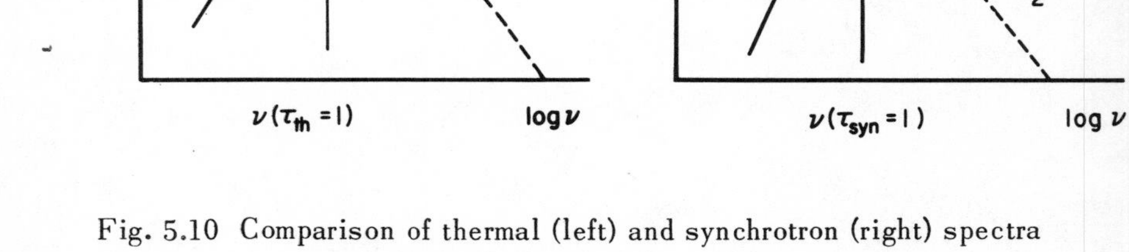

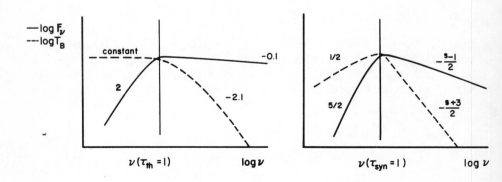

Fig. 5.10 Comparison of thermal (left) and synchrotron (right) spectra

It can be seen that ν_1 and ν_m do not coincide. Their ratio is $\nu_1/\nu_m = \tau_m^{2/s+4}$.

The polarization of the synchrotron source with an isotropic distribution of electrons is predominantly linear; the circular contributions from electrons with positive and negative ψ almost cancel each other. Only if the distribution is highly anisotropic (the anisotropy being significant over the small angle ψ) will there be more ellipticity in the synchrotron radiation. For an isotropic power-law distribution of electrons with an index s in a uniform magnetic field the degree of lienar polarization is, for an optically thin source,

$$\pi = \frac{s+1}{s+7/3} \quad \text{(polarization perpendicular to magnetic field);} \tag{5.42}$$

and, for an optically thin source,

$$\pi = \frac{3}{6s+13} \quad \text{(polarization parallel to magnetic field).} \tag{5.43}$$

We have:

s	2	3	4	5	6
π (thin)	0.69	0.75	0.79	0.82	0.84
π (thick)	0.12	0.10	0.08	0.07	0.06 .

5.4 Comparison of Thermal and Synchrotron Spectra

The solution of the transfer equation for a homogeneous source (cf. Section 1.4)

$$I_\nu = S_\nu \, (1-e^{-\tau_\nu})$$

can be approximated for small τ_ν (optically thin, $e^{-\tau_\nu} \simeq 1 - \tau_\nu$) by

$$I_\nu \simeq S_\nu \cdot \tau_\nu$$

and for large τ_ν (optically thick, $e^{-\tau_\nu} \simeq 0$) by

$$I_\nu \simeq S_\nu .$$

For thermal radiation:

$$\begin{aligned}
&\varepsilon_\nu \propto \nu^{-0.1}, \\
&\kappa_\nu \propto \nu^{-2.1}, \\
&S_\nu = B_\nu(T) \propto \nu^2, \\
&F_\nu(\text{thin}) \propto \nu^{-0.1}, \\
&F_\nu(\text{thick}) \propto \nu^2.
\end{aligned} \tag{5.44}$$

The radiation is unpolarized and isotropic.

For synchrotron radiation (with a power-law electron energy distribution with an index s):

$$\begin{aligned}
&\varepsilon_\nu \propto \nu^{-(s-1)/2}, \\
&\kappa_\nu \propto \nu^{-(s+4)/2}, \\
&S_\nu \propto \nu^{5/2}, \\
&F_\nu(\text{thin}) \propto \nu^{-\alpha} \qquad \alpha = \frac{s-1}{2}, \\
&F_\nu(\text{thick}) \propto \nu^{5/2}.
\end{aligned} \tag{5.45}$$

The radiation is linearly polarized, up to $(s + 1)/(s + 7/3) \sim 72\%(s = 2.5)$ (thin), $3/(6s + 13) \sim 10\%$ (thick), and directional.

Thermal and synchrotron spectra are represented in Figure 5.10. In this figure T_B is the brightness temperature, which is often used in place of the intensity to characterize the radio radiation. It is defined by

$$T_B = \frac{c^2 I_\nu}{2\kappa\nu^2} \propto \frac{I_\nu}{\nu^2} \qquad (5.46)$$

where κ is the Boltzmann constant.

5.5 Energy Losses of Relativistic Electroncs and Synchrotron Spectra

Relativistic electrons, which are present in a synchrotron source, may suffer losses of energy due to a number of processes, like losses from ionization of the surrounding medium, losses due to free-free radiation (bremsstrahlung), inverse Compton scattering, and synchrotron radiation itself. Since those losses are energy dependent, they will affect the energy spectrum of radiating electrons.

Energy losses of relativistic electrons caused by the ionization of the surrounding medium can be considered independent of the electron energy. The losses due to the creation of photons through interactions with the nuclei of the surrounding medium, free-free radiation (brehmsstrahlung) losses, are roughly proportional to the electron enegy. The losses of energy suffered through both synchrotron radiation and inverse Compton scattering are proportional to the square of the electron energy. Of the last two, inverse Compton losses are important in compact radio sources of high brightness, but in extended sources are negligible as compared with synchrotron losses. The relative importance of the ionization losses, free-free radiation losses and synchrotron losses is represented in Figure 5.11.

Let us consider as an example a synchrotron spectrum of a stationary source in which the electron energy losses are balanced by a continuous injection of a power-law distribution of electrons $q_o E^{-s}$:

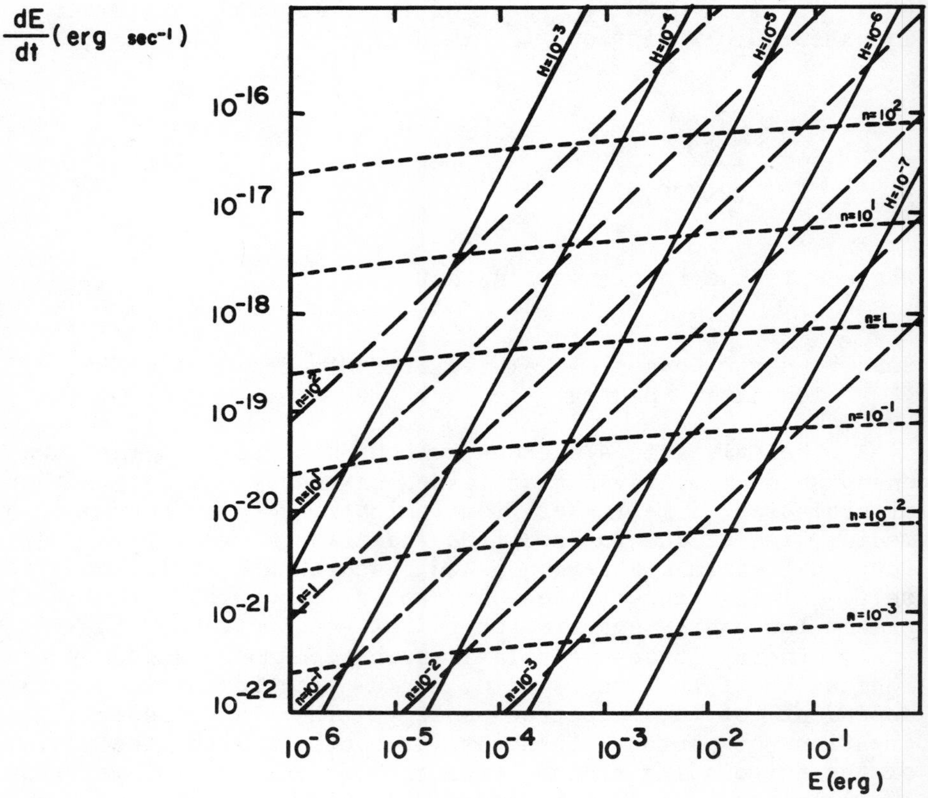

Fig. 5.11 Electron energy loss rate as a function of electron energy in a region containing neutral hydrogen of density n electrons per cubic centimeter. A point line indicates ionisation losses, a broken line free-free radiation losses, and a solid line represents synchrotron radiation losses

$$\frac{\partial}{\partial E}\,[N(E)\varphi(E)] = q_o E^{-s} \ , \tag{5.47}$$

where $\varphi(E)$ represents all the above mentioned loss processes;

$$\varphi(E) = -\,\zeta - \eta E - (\xi_s + \xi_c)E^2 \tag{5.48}$$

ionization losses	free-free radiation losses	synchrotron losses	inverse Compton losses

We have

$$N(E) = q_o \varphi^{-1} \int E^{-s} dE = \frac{q_o}{s-1}\, E^{-s}\left[\frac{\zeta}{E} + \eta + (\zeta_s + \zeta_c)E\right]^{-1} . \tag{5.49}$$

We see that for the region of energies in which the losses independent of energy predominate, $N(e) \propto E^{-(s-1)}$. For intermediate energies $[N(E) \propto E^{-s}]$ the exponent in the electron distribution function remains the same as for the injected particles q(E). Finally, for large energies the losses proportional to the square of the electron energy are the most important and the electron distribution function becomes more steep: $N(E) \propto E^{-(s+1)}$. The situation is represented in Figure 5.12. In the same figure the synchrotron radiation spectrum (broken line) corresponding to the distribution function (solid line) is shown. It is evident from the figure that at low frequencies ionization losses are predominant, free-free radiation losses are most important at intermediate frequencies, and at high frequencies synchrotron losses (and eventually inverse Compton losses) are the most important.

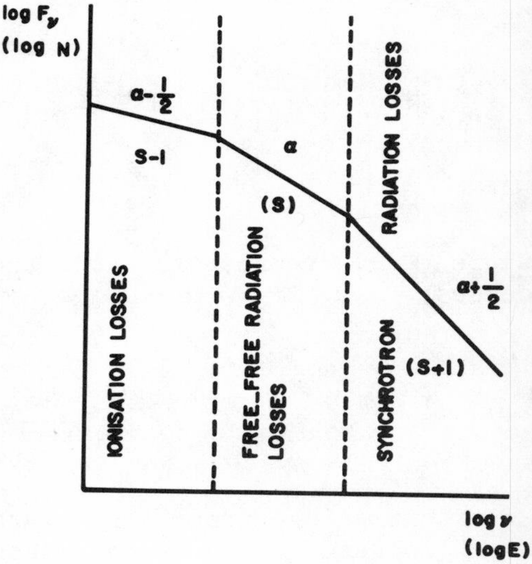

Fig. 5.12
The steady-state distribution function of electrons, N(E), and their synchrotron spectrum, F(ν)

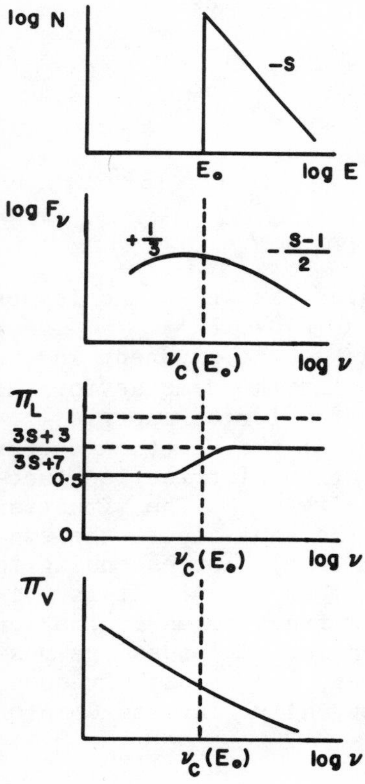

Fig. 5.13
Electron distribution N(E) with a low frequency cut-off at E_o and its synchrotron spectrum F(ν) and polarizations $\Pi_L(\nu)$ and $\Pi_V(\nu)$

5.6 Synchrotron Spectrum at Low Frequencies - Part One

In this section we will discuss possible causes of the low frequency turnover frequently observed in the spectra of radio sources. The term low frequency means in this context that the turnover occurs at frequencies below the range in which the spectrum has a power-law form, and does not necessarily imply that the turnover occurs at low radio frequencies; in fact many compact sources exhibit this turnover at frequencies in the gigahertz range. We will assume that relativistic electrons have an isotroptic power-law distribution of energy in a specified range of energies, and that the magnetic field is uniform.

(1) If the energy distribution of relativistic electrons has a low energy cut-off at E_o , the resulting frequency spectrum of synchrotron radiation of an optically thin source will have a maximum at a frequency $\nu_{max} = 0.3\ \nu_c$ equation (5.29) where ν_c is the critical frequency corresponding to energy E_o . Below ν_{max} the slope of the spectrum will be 1/3, since we will be seeing the low frequency tail of radiation of electrons of energy E_o and higher (see equation 5.32). The linear polarization will vary from the value $(s+1)/(s+\frac{7}{3})$ at frequencies above ν_{max} (equation 5.45) to the value 1/2 at low frequencies. The situation is represented in Figure 5.13.

(2) At sufficiently low frequencies, the cyclotron turnover may be observed in an optically thin synchrotron source (see Figure 5.7). The turnover frequency is $\nu_G/\sin\vartheta$, below it the slope of the spectrum is equal to the energy distribution index s; high energy electrons contribute to both high and low frequency ends of the spectrum. The linear polarization is $(s+1)/(s+7/3)$ at the high end and 1/2 at the low end. Circular polarization varies as $\nu^{-\frac{1}{2}}$ at the high end of the spectrum increasing considerably around the maximum where low energy electrons contribute to the radiation, and decreases below the maximum. Cyclotron turnover is illustrated in Figure 5.14.

(3) Another possible cause of a low frequency turnover can be synchrotron self-absorption occurring in the source. The spectrum, represented in Figure 5.10, has a maximum at the frequency ν_m given by equation (5.41). This frequnecy is of the order of ν_1, frequency at which the optical depth for synchrotron process is unity, which through equations (5.42) and (5.36) can be related to parameters characterizing the source:

$$\nu_1^{\frac{s+4}{2}} = (2c_1)^{5/2} \frac{c_6}{c_5} H_\perp^{1/2} \frac{F_\nu \nu^{\frac{s-1}{2}}}{\Omega}$$

$$= c_{14}(S) H_\perp^{1/2} \frac{F_\nu \nu^{\frac{s-1}{2}}}{\Omega} , \quad (5.50)$$

where F_ν is the flux at a frequency ν at which the source is optically thin, Ω is the angular size of the source in steradians and c_{14} is given in Table 5.1. Below the turnover the flux varies like $\nu^{5/2}$ (equation 5.45). If the source has a significant red shift z, a factor $(1+z)^{1/2}$ should appear on the right hand side of equation (5.50). If Ω is measurable for a given source, the equation (5.50) permits one to estimate the magnetic field. The linear polarization varies from the value of $(s+1)/(s+\frac{7}{3})$ at high frequencies and direction perpendicular to the field, decreasing to zero at a frequency ν_Q related to ν_m by

$$\frac{\nu_Q}{\nu_m} = \left(\frac{\tau_m}{\tau_Q}\right)^{\frac{2}{s+4}} , \quad (5.51)$$

τ_m and τ_Q are the optical depths at maximum flux, i.e. at ν_m (equation 5.44), and ν_Q, respectively and then increasing to the value of $3/(6s+13)$ at low frequencies where the source is optically thick; the plane of polarization at low frequencies is parallel to the direction of the magnetic field. Circular polarization varies as $\nu^{-\frac{1}{2}}$ at high frequencies, reaches a maximum then goes through zero at ν_V,

$$\frac{\nu_V}{\nu_m} = \left(\frac{\tau_m}{\tau_V} \right)^{\frac{2}{s+4}} , \quad (5.52)$$

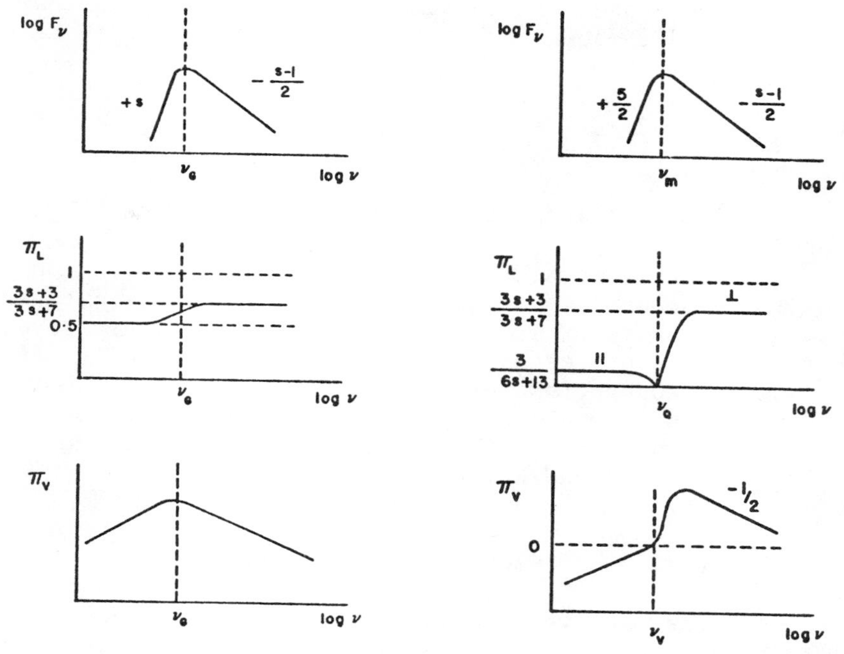

Fig. 5.14 (left) Synchrotron spectrum showing a cyclotron turnover
and the corresponding polarizations
Fig. 5.15 (right) Synchrotron spectrum showing a self-absorption
turnover and the corresponding polarizations

$$\omega_2 = \frac{\omega_B}{2} + \sqrt{\left(\frac{\omega_B}{2}\right)^2 + \omega_0^2}$$

$$\omega_3 = \sqrt{\omega_0^2 + \omega_0^2}$$

$$\omega_1 = -\frac{\omega_B}{2} + \sqrt{\left(\frac{\omega_B}{2}\right) + \omega_0^2}$$

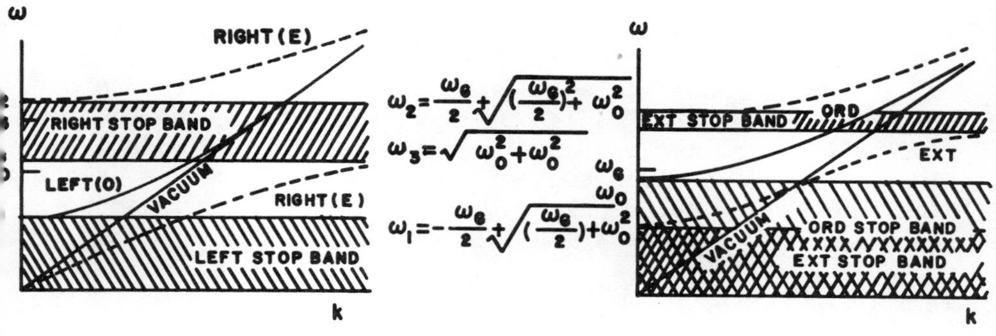

Fig. 5.16 Modes of high-frequency wave propagation in a collisionless
cold plasma in a magnetic field

changing the sense of rotation and decreasing further as $-\nu^{-1/2}$ at low frequencies. τ_V is the optical depth at ν_V,

s	2	3	4	5
τ_Q	6.2	8.2	10.4	12.7
τ_V	4.1	6.2	7.9	9.5

Figure 5.15 illustrates spectrum and polarization of a self-absorbed source.

The three mechanisms described above do not constitute all mechanisms capable of producing a low frequency turnover in the spectra of radio sources. There are several other possible processes, such as the Razin-Tsitovič effect and thermal absorption, either within the source or between the source and the observer, which can modify the spectrum. They will be discussed following the section concerned with the propagation of radiation in a plasma.

5.7. Propagation of High Frequency Waves in a Plasma

A magnetoactive plasma - an electrically neutral (on the average) system of charged and neutral particles interacting with each other mainly through the Coulomb interactions, and interacting with an external magnetic field - affects the propagation of electromagnetic radiation because of the interaction between the charged plasma particles and the field of the wave. If the plasma is cold (that is if thermal velocities of electrons are assumed zero) and collisionless (that is if the frequency of collisions between the particles constituting the plasma is neglected in comparison with the frequency of the wave), and if the velocities of ions are neglected as compared with those of electrons, there are two modes of propagation of electromagnetic radiation and a third nondispersive oscillating mode. The last one is a longitudinal oscillation of charges in the Coulomb field occurring with a characteristic frequency, called plasma frequency ν_o,

$$\nu_o = \frac{\omega_o}{2\pi} = \frac{1}{2\pi}\sqrt{4\pi\frac{e^2}{m}N} = 8920\ N^{1/2}\ , \tag{5.53}$$

where N is the number density of thermal electrons. Plasma oscillations do not propagate in the approximation of a cold plasma; beyond this approximation in a warm plasma (non-zero thermal but non-relativistic velocities v_{th} of electrons) the dispersion relation for plasma oscillation is

$$n^2 = \frac{c^2}{v_{th}^2}\left(1 - \frac{\omega_o^2}{\omega^2}\right), \tag{5.54}$$

where

$$n = \frac{kc}{\omega} \tag{5.55}$$

is the index of refraction (k is the wave number). Plasma oscillations have therefore a non-zero group velocity in a warm plasma and can propagate at frequencies ω larger than the plasma frequency ω_o.

In what follows we will be interested in propagation of the two transverse electromagnetic modes in the approximation of a cold plasma. The indices of refraction of the two modes called <u>ordinary mode</u> (plus sign) and <u>extraordinary mode</u> (minus sign) are given by

$$n^2 = 1 - \frac{v}{\dfrac{1-u_T/2}{1-v} \pm \sqrt{\dfrac{u_T^2}{4(1-v)^2} + u_L}} \tag{5.56}$$

where

$$v = \frac{\omega_o^2}{\omega^2}, \qquad u_L = \frac{\omega_G^2}{\omega^2}\cos^2\vartheta, \qquad u_T = \frac{\omega_G^2}{\omega^2}\sin^2\vartheta,$$

ϑ is the angle between the direction of wave propagation and that of a uniform external magnetic field (ω_G is the gyrofrequency of electrons in that field).

If the following inequality is satisfied

$$\left(\frac{\omega_G}{\omega}\right)^2 \frac{\sin^4\vartheta}{4\cos^2\vartheta} << \left(1 - \frac{\omega_0^2}{\omega^2}\right)^2 , \tag{5.57}$$

equations (5.56) simplify considerably yielding

$$n^2 \cong 1 - \frac{\omega_0^2}{\omega(\omega \pm \omega_G \sin\vartheta)} . \tag{5.58}$$

The approximation (5.57), called quasilongitudinal approximation, holds for a broad range of angles ϑ if $\omega_G/\omega << 1$ and $\omega_0/\omega << 1$, which is often the case in astrophysical applications. The two modes are circularly polarized. For $\vartheta = 0$ we have longitudinal propagation.

If an inequality opposite to that of (5.57) is satisfied, we have the quasitransverse approximation in which the two modes are linearly polarized, and

$$n^2_{ord} \cong \frac{1 - v}{1 + (1-v)\cot^2\vartheta} , \tag{5.59}$$

$$n^2_{ext} \cong \frac{1-v}{1 - \frac{u_T}{1-v}} . \tag{5.60}$$

When $\vartheta = 90°$ we have transverse propagation.

The properties of the modes are illustrated in Figure 5.16 in the case of longitudinal and transverse propagation. There are several singular points on the diagrams. At ω_0, ω_1, and ω_2, called cut-offs, $k^2 \to 0$. At ω_G and $\sqrt{(\omega_0^2 + \omega_G^2)}$, called resonances, $k^2 \to \infty$. If the wave propagates in a variable density plasma and approaches places where $k^2 \to 0$ and $k^2 \to \infty$, it will experience total reflection and strong absorption, respectively. The resonances are experienced by the extraordinary wave. In the case of longitudinal propagation the resonance is due to interaction of electron on a cyclotron orbit with the (right) circularly polarized wave, this occurs at ω_G. In the case of transverse propagation the resonance is due to excitation of longitudinal oscillations when electric vector of wave points in the direction of propagation during part of the cycle. The resonance occurs at $\sqrt{(\omega_o^2+\omega_G^2)}$.

When collisions are taken into account in a cold plasma, the refractive index is not infinite at resonances, but passes through unity; and there is cyclotron absorption (a process reverse to cyclotron emission) around resonances. In warm plasmas, that is when the electron temperature is not assumed zero, but the motion of electrons is still assumed nonrelativistic, the longitudinal and the ordinary modes are as in a cold plasma in the case of longitudinal propagation. The extraordinary mode has different properties around resonance than in a cold plasma: n^2 is not infinite but goes through unity, and there is a cyclotron damping with a maximum at ω_G (Figure 5.17). There are no resonances at harmonics of ω_G. When propagation is transverse, the ordinary mode is essentially unchanged, the extraordinary mode is partially longitudinal and coupled with the longitudinal mode, which is partially transverse. Extraordinary wave has resonance dispersion and exhibits harmonic structure.

In many applications to galactic and extragalactic astronomy ω_G^2/ω^2 is very small (e.g., for $H = 10^{-5}$ gauss, $\omega_G = 200$ Hz), and ω_o^2/ω^2 is also small because of the low electron densities. Under these circumstances the condition (5.57) is satisfied for a cold plasma for a wide range of angles ϑ. Moreover, since ω_G is negligible in comparison with ω, there is no need to differentiate between the ordinary and extraordinary modes; the dispersion equation for either of these modes [equations (5.58)] becomes

$$n^2 \cong 1 - \frac{\omega_o^2}{\omega^2} . \qquad (5.61)$$

The plasma can therefore be treated as isotropic except in problems involving the computation of the phase difference between the ordinary and extraordinary rays. An important example of this exception is the problem of the Faraday

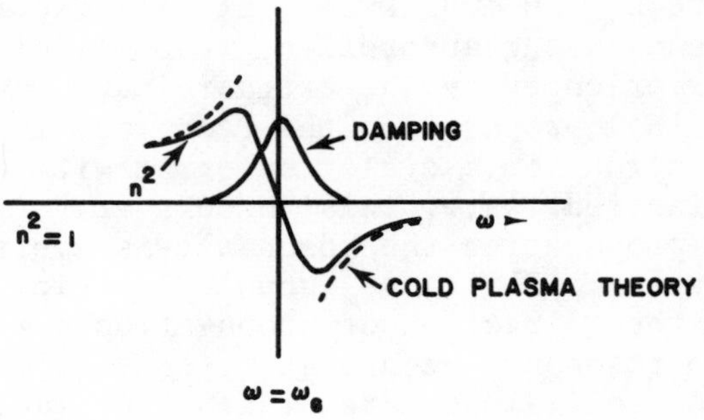

Fig. 5.17 Behavior of the refractive index of the warm plasma in the neighborhood of a resonance

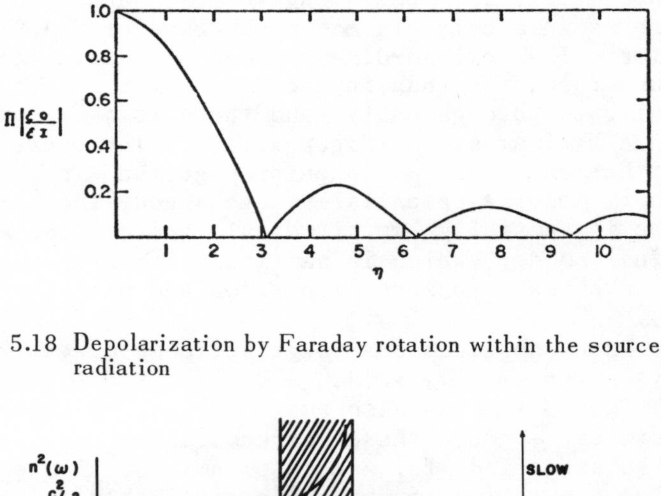

Fig. 5.18 Depolarization by Faraday rotation within the source of radiation

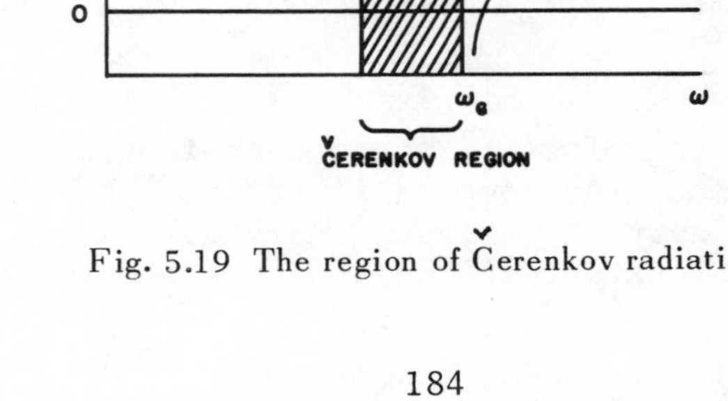

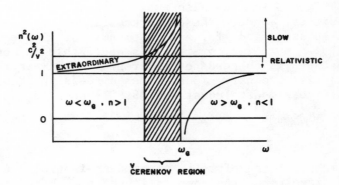

Fig. 5.19 The region of Čerenkov radiation

rotation of the plane of polarization of an electromagnetic wave passing through a slab of plasma. The difference of phase ψ, $(\omega/c) \cdot (n_{ord}-n_{ext})s$, is proportional to the path length s through the plasma, which in astronomical conditions is very large and thus offsets the very small difference in the refractive indices. The Faraday rotation angle χ_F in radians is

$$\chi_F = \psi/2 = 1/2 \frac{\omega}{c}(n_{ord} - n_{ext})\, s$$

$$= 1/2 \frac{\omega_o^2}{c} \frac{\omega_G \cos\vartheta}{\omega^2 - \omega_G^2 \sin\vartheta} s \simeq 1/2 \frac{\omega_o^2 \omega_G \cos\vartheta}{c\omega^2} s \tag{5.62}$$

$$= 0.93 \times 10^6 \frac{NH_{\parallel}\, s}{\omega^2} .$$

When $NH_{\parallel}$ varies along the path, $NH_{\parallel}\, s$ should be replaced by $\int NH_{\parallel}\, ds$. The plane of polarization of radiation passing through a slab s of plasma will therefore be rotated by an angle χ_F. This angle is a function of ω^2, therefore measuring the position angle of the polarization of a source at several frequencies permits one to determine the quantity $\int NH_{\parallel}\, ds$, called Faraday rotation measure, characterizing the slab of plasma.

It should be noted that the plane of polarization of the radiation emitted at various depths within a source containing plasma will be rotated by various amounts. Therefore, even if the source is homogeneous and the emitted polarization identical throughout a source, there will be depolarization of radiation emerging from the source due to the differential rotation. For an optically thin source, the degree of polarization of a homogeneous source will be (Figure 5.18)

$$\pi = \frac{s+1}{s+7/3} \cdot \left| \frac{\sin \tilde{\beta} s}{\tilde{\beta} s} \right| , \tag{5.63}$$

where $\tilde{\beta}$ is the rate of the Faraday rotation and is given by

$$\tilde{\beta} = \frac{d\chi_F}{ds} . \tag{5.64}$$

Let us conclude this section with a digression concerning Čerenkov radiation. The Čerenkov condition (equation 5.2) can be written as

$$v > u = \frac{c}{n}, \qquad n > \frac{c}{v} > 1 \tag{5.65}$$

that is, refractive index must be larger than unity. For a collisionless isotropic electron plasmas without magnetic field

$$n^2 = 1 - \frac{\omega_o^2}{\omega^2} < 1 \tag{5.66}$$

and Čerenkov radiation in the electromagnetic mode is impossible. It is however possible to induce plasma oscillations by fast electrons through Čerenkov process. In a magnetoactive collisionless plasmas Čerenkov radiation is possible for the extraordinary wave (minus sign) (see Figure 5.19). If the electron velocity is small (nonrelativistic) the region of Čerenkov emission is narrow and close to ω_G, and it closely resembles cyclotron radiation.

5.8 Synchrotron Spectrum at Low Frequencies - Part Two

If within the radio source, or on the way between the source and the observer there is a substantial amount of cold plasma, the synchrotron spectrum of the source will be modified by thermal absorption in this plasma.

(4) If the plasma is within the source (internal thermal absorption), the solution of the transfer equation (for a source optically thin against synchrotron absorption) is

$$I \propto \nu^{\frac{5-s}{2}} (1 - e^{-\kappa_o l \nu^{-2}}) , \tag{5.67}$$

since the thermal absorption coefficient $\kappa \simeq \kappa_o \nu^{-2}$ in the radio region. At high frequencies $I \propto \nu^{1-s/2}$, same as without plasma. At low frequencies $I \propto \nu^{(5-s)/2}$. The maximum in the spectrum occurs at

$$\nu_T = \frac{\nu_1^{(T)}}{\sqrt{\tau_T}} \ , \qquad (5.68)$$

where $\nu_1^{(T)}$ is a frequency at which the optical depth (for thermal absorption) is unity,

$$\nu_1^{(T)} \cong 10^3 \ N_e\sqrt{l} \qquad (5.69)$$

(N_e is the number density of electrons and r the linear size of the source, the assumed temperature of the electrons is 10^4 °K), and τ_T is the optical depth at ν_T,

s	1	2	3	4	5	
τ_T	0	0.55	1.26	2.34	∞ .	(5.70)

The linear polarization of synchrotron radiation will be modified by the presence of thermal electrons and the degree of polarization will exhibit a pattern characteristic of depolarization due to Faraday rotation (equation 5.63). Internal thermal absorption turnover is illustrated in Figure 5.20.

(5) The effect of <u>external thermal absorption</u> is described by an exponential cut-off of a synchrotron power-law spectrum pronounced at low frequencies:

$$I \propto \nu^{-\frac{s-1}{2}} e^{-(\nu_1^T/\nu)^2} \ , \qquad (5.71)$$

due to the passage of radiation through a plasma region outside the source. The frequency ν_T of spectral maximum is given by equation (5.68), where now τ_T = (s-1)/4,

s	1	2	3	4	5
τ_T	0	0.25	0.50	0.75	1.0

The degree of polarization, equal to (s + 1)/(s + 7/3) re-

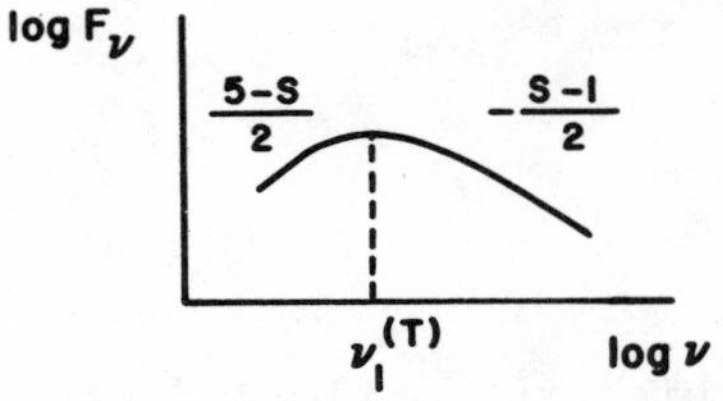

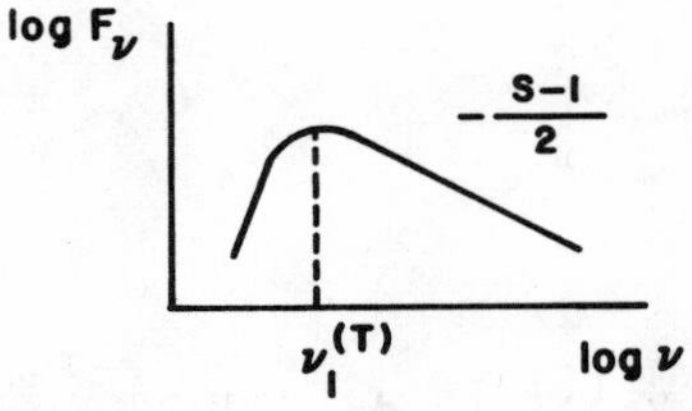

Fig. 5.20 (left) Internal thermal absorption turnover in a spectrum of a synchrotron source

Fig. 5.21 (right) External thermal absorption turnover in a spectrum of a synchrotron source

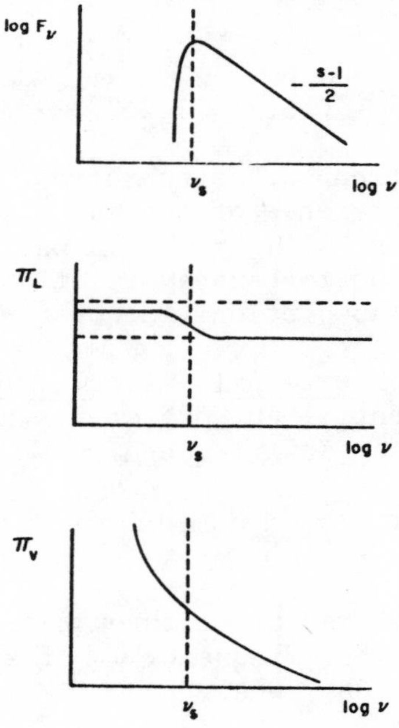

Fig. 5.22 Synchrotron spectrum showing the Razin-Tsitovič effect and the corresponding polarizations

mains unchanged, while the position angle of the polarization vector undergoes Faraday rotation (equation 5.62). External thermal absorption turnover is illustrated in Figure 5.21.

(6) The plasma within a synchrotron source affects the process of emission at low frequencies where the departure of the index of refraction of the plasma from unity significantly changes the beam-width of the radiation emitted by an electron. The beam-width in a vacuum is of the order of mc^2/E, as we remember from Section 5.3, meanwhile in a plasma it is of the order of $\sqrt{(1 - n^2)}$ at frequencies such that

$$1 - n^2(\omega) > (mc^2/E)^2 , \tag{5.72}$$

or since $\omega \simeq \omega_c \simeq (\frac{E}{mc^2})^2 \omega_G$ (equation 5.22), at frequencies for which

$$1 - n^2(\omega) > \frac{\omega_G}{\omega} . \tag{5.73}$$

We can determine the frequency ν_s below which this effect, called Razin-Tsitovič effect, exponentially cuts down the intensity of synchrotron radiation, using the value of the refractive index for a cold plasma (assuming $\omega_G << \omega_o$):

$$\nu_s \cong 20 \frac{N_e}{H} . \tag{5.74}$$

The process affects radiation of all polarizations in the same way and therefore the polarization of the radiation is only slightly modified by the Razin-Tsitovič effect. In addition, the radiation may exhibit depolarization due to Faraday effect of cold plasma.

If only relativistic self-plasma is present in the emitting region, the frequency ν_s is substantially lower

$$\nu_s \cong 20 \frac{s-1}{s} \frac{N_r}{H} \frac{mc^2}{E_{min}} , \tag{5.75}$$

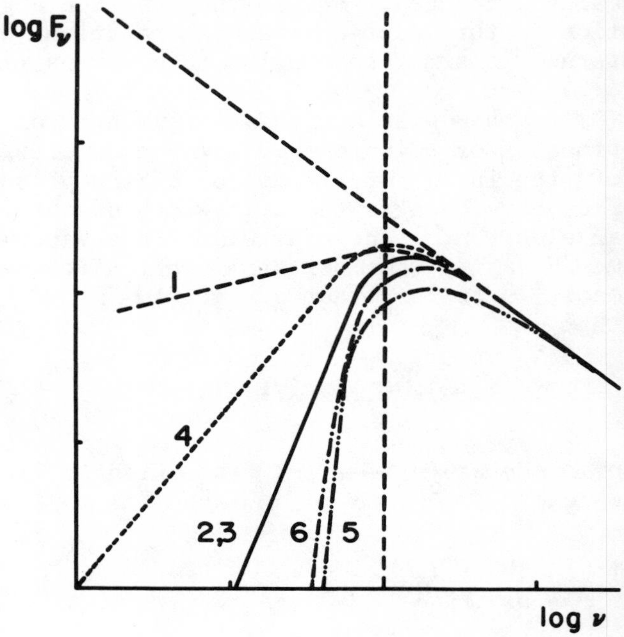

Fig. 5.23
A summary of processes leading to low frequency turnovers in synchrotron spectra

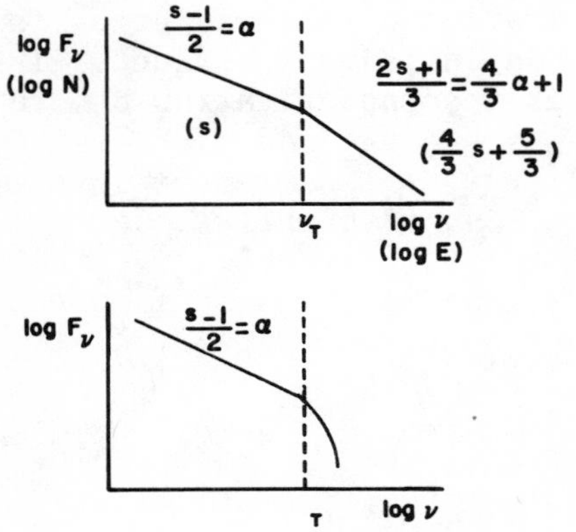

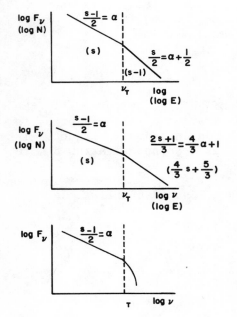

Fig. 5.24
High frequency shape of synchrotron spectra

where N_r is the density of relativistic electrons and the energy E_{min} refers to the lower end of the electron distribution. The polarization will be somewhat affected by the Razin-Tsitovič effect alone, but the presence of the relativistic plasma can affect propagation of radiation and significantly alter the polarization if the density of electrons is sufficiently high. At high frequencies depolarization is similar to that due to cold plasma, but less efficient by a factor of the order of mc^2/E_{min}. At low frequencies (that is, frequencies of the order of the critical frequency corresponding to the energy E_{min}) the propagation modes are linearly polarized rather then circularly polarized, and conversion of linear polarization into circular (circular repolarization) may be of importance. The Razin-Tsitovič effect is illustrated in Figure 5.22.

An analysis of the low frequency turnover in the spectrum of a radio source and of its polarization in terms of one or more of the processes discussed here is a potential source of information about parameters characterizing physical conditions in the source. A summary of processes leading to low frequency turnovers in the spectra of radio sources is given in Figure 5.23.

5.9 Synchrotron Spectrum at High Frequencies

The shape of the high frequency part of a synchrotron spectrum, which is emitted by an initially power-law energy distribution of electrons, depends mainly on the balance of the rate of synchrotron and (for compact radio sources) inverse Compton scattering losses, and of the rate of replenishment (injection) of electrons in the radiating region. Those two loss mechanisms are of particular interest at high frequencies (since their rates increase with the square of the electron energy) where they produce curvature in an originally straight power-law spectrum.

Figure 5.24 represents the high frequency shape of a synchrotron spectrum from an initially isotropic distribution of electrons (it should be remembered that synchrotron energy losses depend on the electron pitch angle) (1) with no replenishment of electrons in the source during the last t seconds and (2) with continuous steady injection of electrons characterized by a power-law distribution of energies. The curvatures in the spectra manifest themselves at a frequency

$$\nu_t = \frac{c_1}{c_2^2 H^3 t^2} = 1.12 \times 10^{24} H^{-3} t^{-2} . \qquad (5.76)$$

When no replenishment occurs, t is the time which has elapsed since the termination of the injection of relativistic electrons into the radio source, when the injection is continuous t stands for the time interval since it began. In this second case the curvature does not depend on the injection rate, but the height of the spectrum does.

The third curve on Figure 5.24 illustrates the shape of a power-law synchrotron spectrum with a high energy cut-off in the distribution of electrons, the cut-off occurring at energy E_o ; the frequency ν_t will be, in this case, a function of this energy through the known relationship for the critical frequency:

$$\nu_t = c_1 H E_o^2 . \qquad (5.77)$$

The high frequency curvature in the spectra of radio sources is a potential source of information about the character of the injection process and about ν_t, i.e. about the timescale t if the magnetic field H within the source is known.

The above discussion referred to a stationary source. If a source is expanding, we must take into account that the rates of energy losses due to the described mechanisms will vary significantly with the radius R of the expanding source. Assuming that the magnetic flux remains conserved during the expansion of the source, the rate of synchrotron losses varies as R^{-4} E^2 , the rate of Compton losses varies as R^{-8} E^2 , the rate of free-free radiation losses varies as R^{-3} E, and the rate of ionization losses varies as R^{-3}. To these we must add the energy losses due to the expansion of the source (their rate varies as R^{-1} E). The energy gains by the Fermi statistical acceleration process depend on $R^{-5/3}$ E. If the energy of every relativistic particle decreases as R^{-1} during the expansion of the synchrotron-emitting region, we can see that the rates of energy losses and gains will depend on various powers of the radius, from minus two for expansion losses to minus ten for inverse Compton effect losses. Therefore during an evolution of the component of a source the relative contribution of different processes may also undergo changes. All of this will, of course, have to be taken into account when discussing the evolution of spectra of expanding components. Some of the sources like quasars show changes of the flux in time at different frequencies. The general character of these variations can be roughly described in terms of a model of an expanding source which is optically thin at higher frequencies and thick at lower frequencies.

6. Physics and Astronomy of Celestial X-Ray Sources

6.1. Introduction

The term "X-ray astronomy", unlike, for example, "radiative transfer" or "cosmology", does not denote a well-defined subfield of theoretical astrophysics. Instead it refers to <u>observations</u> in a particular spectral band, roughly 10^{17} - 10^{21} Hz. These observations tend to reveal processes involving particle energies comparable to the photon energies, i.e. $\gtrsim$ 1 keV, which if thermal correspond to temperatures $T \gtrsim 10^7$ °K. Thus through X-ray observations we encounter physical conditions very different from those in, say, gaseous nebulae, or at the surfaces of stars. We therefore require bits and pieces of classical and atomic physics which, until the advent of X-ray astronomy, were frequently not part of a fashionable astronomer's experience. This chapter is an attempt to introduce the student to some of these pertinent results and the ways in which they are applied to X-ray sources. We have space to survey only the most important topics, but we have tried to provide useful references to lead the interested student deeper into some of the problems.

X-ray astronomy is a young and rapidly developing field. Its relative youth is due principally to the difficulties and limitations of observation. At photon energies $\gtrsim$ 20 keV, celestial X-rays can be detected, weakly, at

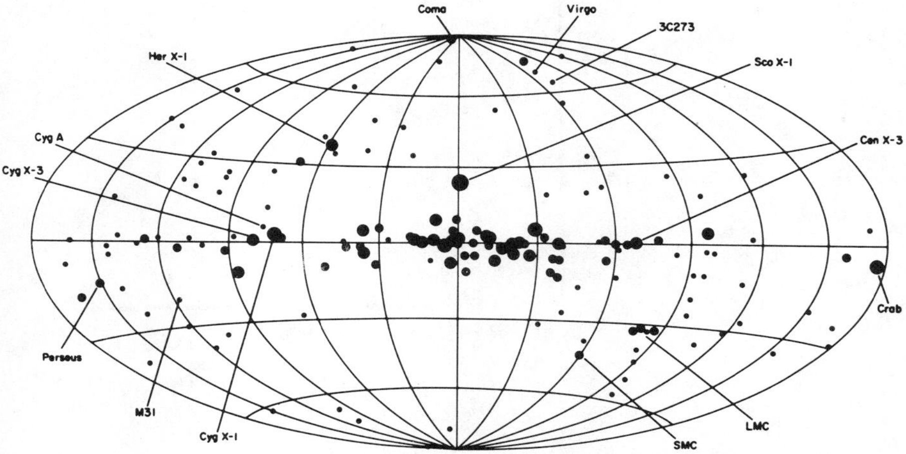

Fig. 6.1 (by permission of H. Gursky). The 161 sources seen by Uhuru. The map is an equal-area projection in galactic coordinates; the center point is the direction to the galactic center, and the north galactic pole is at the top. The size of each dot is proportional to the apparent intensity of the X-ray source.

balloon altitudes, but below this energy the atmosphere is so opaque that observations can be made only with rockets or space vehicles. At this writing (February 1975), more than half our information on celestial X-ray sources (apart from the sun, which we do not discuss here) has come from one space vehicle, the SAS-1 or "Uhuru" satellite, launched in December 1970. Theory and observation are growing up together. Few or no important sources have been predicted successfully except in the vaguest way, and existing "theories" are only post facto attempts to organize and rationalize the observed facts. In this melange, rapid change must be expected, and it is possible to keep up with the field only by frequent inspection of the journals. Three fairly recent compendia (Bradt and Giacconi 1973, Hegyi 1973 and Giacconi and Gursky 1974) are useful and give references to earlier work.

Roughly 200 sources are known. The best single listing is the third Uhuru catalog (Giacconi et al. 1974), comprising 161 sources, which appear with numbers such as 3U 1642-45 , this denoting a source with approximate right ascension $16^h\ 42^m$ and declination $-45°$. (Many of the well-known sources have more familiar names also.) The catalog

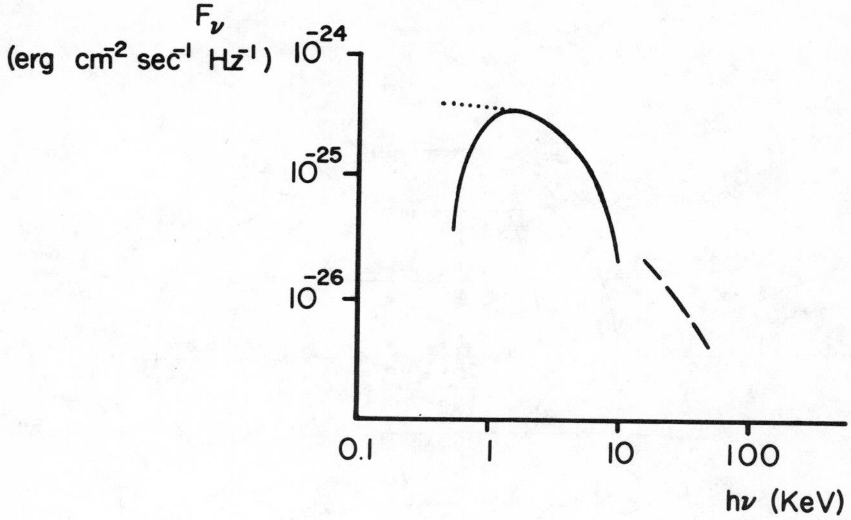

Fig. 6.2 X-ray spectrum of Scorpius X-1 (sketched). The dashed line is a possible hard component ($h\nu>20$ keV). The dotted line is the inferred spectrum at the source, after correction for interstellar X-ray absorption.

contains numerous references. The distribution of these sources in galactic coordinates is shown in Figure 6.1. Their concentration toward the galactic plane shows clearly that many of these objects must be galactic, though some, especially among those at high latitudes, are known to be in external galaxies. In addition to the resolved sources, there seems to be a diffuse, more or less isotropic, X-ray background, not yet resolved, to which we shall return later.

6.2. Scorpius X-1

Avoiding details, especially observational details, as much as possible, we shall discuss several important sources and the theories currently applied to them, in roughly historical order. Scorpius X-1 (3U 1617-15) is the brightest and first-discovered source, apart from the sun. Much of the physics relevant to other sources enters naturally in consideration of this object. Its location at $\ell^{II} \simeq 0°$, $b^{II} \simeq 24°$, well out of the galactic plane, suggests that, if not extragalactic, it might be fairly local, say within a few hundred pc.

The observed flux-density spectrum of its X-rays is

sketched on a log-log plot in Figure 6.2. The possible "hard" (> 20 keV) X-ray component shown does not fit smoothly onto the spectrum observed at lower energies. Observations > 20 keV, made mostly from balloons, are not constant or even frequent. This component may be variable, and even its existence is not very well established.

The best-observed portion of the spectrum, between 1 and 10 keV, is definitely curved on a log-log plot, and looks like thermal bremsstrahlung (free-free) emission by a hot dilute gas cloud. The spectrum of bremsstrahlung radiated by electrons in a Maxwellian velocity distribution, colliding with ions in a plasma having some given distribution of net ionic charges Z, can be calculated quantum-mechanically. Under certain conditions usually applicable in X-ray sources, the result for the net bremsstrahlung emission coefficient of the plasma (Karzas and Latter 1961, Tucker and Gould 1966) is

$$j_{ff}(\nu,T) \simeq 5.44 \times 10^{-39} n_e (\sum_Z n_Z Z^2) \bar{g}(\nu,T) \times T^{-1/2} e^{-\frac{h\nu}{kT}} \text{erg}(\text{cm}^3 \text{sec sr Hz})^{-1}. \quad (6.1)$$

Here n_e (cm^{-3}) is the free electron density and n_Z is the density of ions of net charge Z. Note the ν-dependence. The quantity $\bar{g}(\nu, T)$, called the "temperature-averaged Gaunt factor", is a weak, tabulated function of ν and T, always ~ 1. For a rough approximation we may set it = 1. Then the ν-dependence of j_{ff} is exp $(- h\nu /kT)$, the characteristic exponential shape of thermal bremsstrahlung, very similar to what we see in Figure 6.2.

We shall not discuss the quantum mechanics of this result. It is, however, useful for the student to understand _why_ a thermal (Maxwellian) distribution of electron velocities gives rise to this particular shape, because _non_thermal bremsstrahlung is also discussed frequently. The spectrum of bremsstrahlung produced by electrons having a _single_ kinetic energy E_0 (Figure 6.3; cf. Heitler 1954) is nearly flat for $h\nu < E_0$, and cuts off sharply at E_0, since the accelerating electron cannot radiate more energy than it has. In a thermal distribution, most of the electrons have kinetic energies near kT. Thus the summed spectrum must be roughly flat for $h\nu \ll kT$, and must fall off sharply at $h\nu \sim kT$. Using similar reasoning, one can understand qualita-

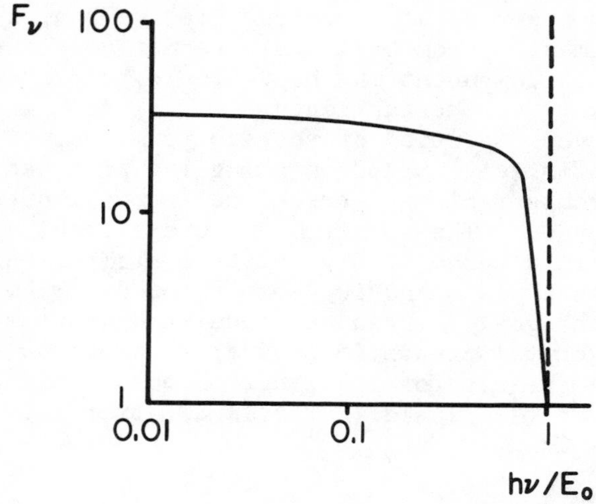

Fig. 6.3 Sketch of a typical spectrum of bremsstrahlung from electrons of a given kinetic energy E_o.

tively how the bremsstrahlung spectrum will differ when the electrons have a nonthermal energy distribution, e.g. some power law.

Now the X-ray spectrum of Sco X-1 is in fact time-variable. The flux between, say, 2 and 10 keV varies irregularly by about a factor of 3, the maximum being ~70 photons $cm^{-2}sec^{-1}$. (For comparison, the weakest 3U source is about ten thousand times weaker.) The best T for a bremsstrahlung fit to the data varies too, by about a factor of 2, a typical value being $T \simeq 6 \times 10^7$ °K ($kT \simeq 5$ keV). A variable optical object has been identified with this source, and there are some correlations between the X-ray and optical variations (Blumenthal and Tucker 1974). Besides an optical continuum, the object shows broad emission lines which suggest that cooler gas ($T \sim 10^4$ °K) is present in addition to the hot gas producing the X-rays.

The distance D to Sco X-1 is extremely uncertain, though it is generally supposed to be galactic. One approach to this, which we may introduce here, is that of interstellar X-ray absorption. Interstellar atoms, if not completely ionized, absorb X-rays by photoelectric effect. The general behavior of the effective cross section σ_e <u>per hydrogen atom</u> (including contributions from heavier species, weighted for their cosmic abundances) for neutral gas is as shown in Figure 6.4 (Brown and Gould 1970; cf. Cruddace *et al.* 1974;

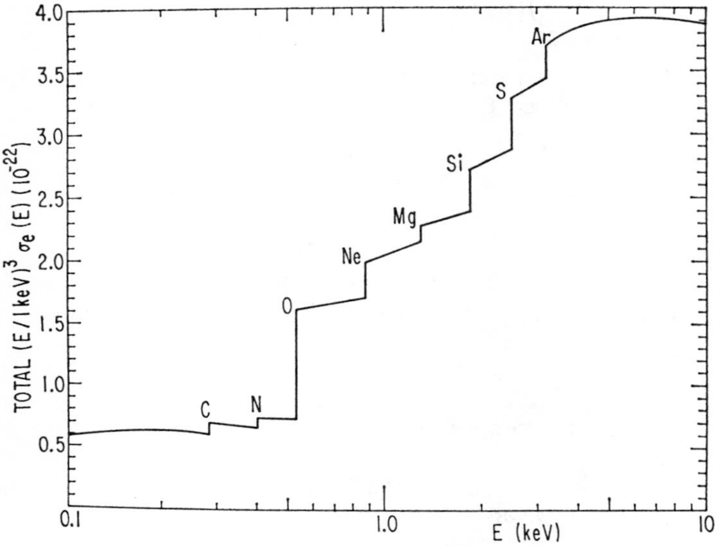

Fig. 6.4 (Brown and Gould 1970). Total photo-ionization cross section per hydrogen atom [$\times(E/1\text{ keV})^3$ in units 10^{-22} cm^2] as a function of incident photon energy. The elements responsible for the jumps due to their respective K edges are indicated.

above 6 keV especially cf. Fireman 1974). The dependence on photon energy is approximately E^{-3} ; this has been taken out in the figure. In addition there are absorption edges corresponding to the K edges of the various elements present. Complications can arise in calculating X-ray absorption when the composition and state of the interstellar medium are unknown. In hot regions, hydrogen and even helium might be fully ionized and would then not absorb, but the K shells of heavier atoms, which produce most of the absorption above 500 eV, would probably not be ionized. Molecules could also affect the cross section σ_e. Eventually X-ray absorption will become a useful probe of the interstellar medium, but at present the energy resolution of soft X-ray detectors is not good enough even to reveal the absorption edges.

Forgetting these complications and assuming that the gas is neutral, we can use Figure 6.4 and say that the optical depth over a path length D is

$$\tau(h\nu) = n_H \sigma_e(h\nu) D = N_H \sigma_e(h\nu) \quad , \qquad (6.2)$$

where N_H is the column density of hydrogen atoms in cm^{-2}. The transmitted flux from an X-ray source is

$$F_\nu(h\nu) = F_{\nu 0}(h\nu) e^{-\tau(h\nu)} \quad , \qquad (6.3)$$

where $F_{\nu 0}$ is the flux density in the absence of absorption. Since $\tau \mathrel{\tilde{\propto}} (h\nu)^{-3}$, we expect a rather sharp onset of absorption at sufficiently low $h\nu$. Using equation (6.2) and Figure 6.4, we find that at $h\nu = 1$ keV, the column density out to unit optical depth is $N_H \simeq 5 \times 10^{21} cm^{-2}$, which at $n_H \sim 0.5 cm^{-3}$ corresponds to $D \sim 3$ kpc. Thus for sources kiloparsecs away, we expect strong absorption at 1 keV and even above. At a few hundred eV, even sources with $D \simeq 200$pc should show absorption.

Such an absorption turnover is in fact observed in Sco X-1 (Bunner et al. 1972). In Figure 6.2 we sketch the shape of the turnover and the inferred (bremsstrahlung) spectrum at the source. The implied column density is $N_H \sim 3 \times 10^{21} cm^{-2}$, which indicates a distance $D \sim 2$ kpc. (This would need re-examination, for at $D \sim 2$ kpc this high-latitude line of sight has reached out of the galactic disk into a region where n_H is small). Recent observations, however (Deerenberg et al. 1973, Moore et al. 1973), suggest that N_H is time-variable and reaches values as low as $6 \times 10^{20} cm^{-2}$. These soft X-ray observations, made from rockets, are rather limited. If the result is correct, the absorption must be mostly in a circumstellar cloud rather than interstellar, and the minimum N_H which occurs then gives a sort of upper limit on the distance D; $6 \times 10^{20} cm^{-2}$ corresponds to around 400 pc. It is of course necessary to compare the derived N_H with column densities to neighboring stars derived by other techniques.

Other astronomical evidence on the distance is contradictory (Wallerstein 1967, Felten and Humphreys 1973). D must be regarded as quite uncertain, though $D \lesssim 400$ pc seems not unlikely at present. The inferred luminosity of Sco X-1 of course depends on D. For $D \simeq 400$ pc, the total luminosity under the bremsstrahlung curve is P_{XR}

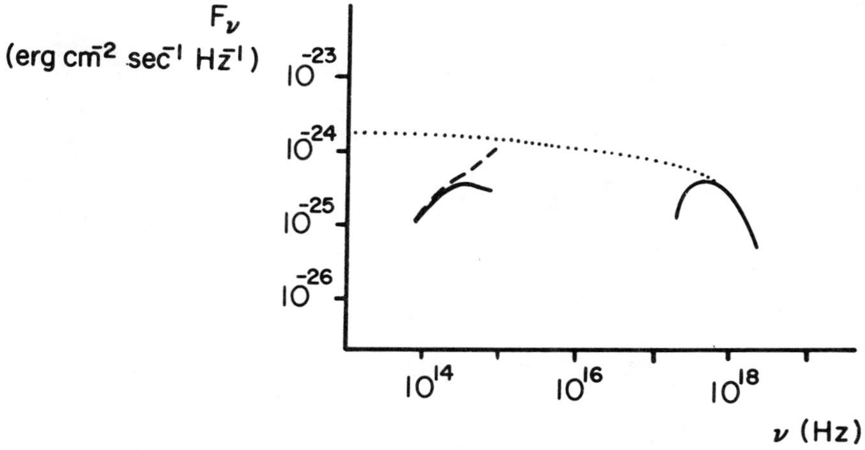

Fig. 6.5 X-ray and optical continua of Sco X-1 (sketched). This is a typical situation ; all fluxes show time variations. Solid lines are observed spectra. The dotted line is the inferred bremsstrahlung spectrum of the X-ray source. The dashed line is the optical continuum corrected for interstellar extinction $A_V \approx 0.7$ mag. It is believed that the optical flux falls below the bremsstrahlung extrapolation because the source is not optically thin at these low frequencies.

$\simeq 1.0 \times 10^{37}$ erg sec^{-1}, or 2×10^{3} $L_\odot$ (The optical luminosity is $\sim 10^{-4} P_{XR}$).

More information about Sco X-1 can be gained by plotting the optical and X-ray continua on a common graph (Figure 6.5). We see that the observed optical continuum falls below the bremsstrahlung extrapolation. It is clear that we should expect this at sufficiently low frequencies. Why? For $kT \simeq 5$keV, we have $h\nu << kT$ at optical ν, so the black-body specific intensity, which is also the surface brightness of a black body, assumes its Rayleigh-Jeans form:

$$B_\nu(\nu) \simeq \frac{2kT\nu^2}{c^2} \text{ erg(cm}^2\text{sec sr Hz)}^{-1}. \qquad (6.4)$$

At sufficiently low ν, depending on the angular size of the source, this must fall below the bremsstrahlung curve. But the thermal gas cloud cannot radiate more than a black body! The actual emission by the cloud must therefore fall below the bremsstrahlung curve at low ν. One physical reason this happens is that the cloud becomes optically thick due to free-free reabsorption, the inverse of bremsstrahlung emission.

The fact that this happens at $\sim 10^{15}$ Hz, rather than elsewhere, permits an estimate of the angular size of the source. On the other hand, the flux density in the X-ray band, where the cloud is optically thin, is proportional to $n^2 VD^{-2}$, where D is the uncertain distance, V is the volume of emitting gas, and n is its density. (The emission coefficient in equation (6.1) is essentially proportional to n^2). From this we can develop a one-parameter family of simple homogeneous spherical models of this gas cloud, with D as parameter. For D $\simeq$ 200 to 1000 pc, one finds $n \sim 10^{16}$ cm^{-3} (denser than the solar corona) and the radius $R \sim 10^9$ cm (about the size of a white dwarf).

Complications arise in these models (Felten and Rees 1972, Illarionov and Sunyaev 1972), leading to difficult problems in radiative transfer. The main opacity in the models over the observed frequency ranges is due not to free-free reabsorption but to Thomson scattering of photons by free electrons. This affects the shape of the infrared and optical spectrum. Furthermore, at $T \simeq 6 \times 10^7$ °K, the mean thermal velocity of the electrons is $\langle v \rangle \simeq 0.18$ c, and so these are really Compton scatterings, with non-negligible frequency shifts, implying energy exchange between photons and electrons. The optical thickness for Compton scattering along a radius R in these models is $\tau_{es} \sim 10$, so that photons emitted in the interior must diffuse outward. En route they are trying to come to thermal equilibrium with the Maxwellian electrons; they escape before reaching true equilibrium, but distortions in the emergent bremsstrahlung spectrum may appear, in particular a bump around 2.70 kT, the mean energy of photons in a black-body distribution. This process is called "Comptonization" by the Soviet writers. The absence of an evident bump of this kind in the X-ray spectrum of Sco X-1 gives an additional constraint on the models. It appears that the effects of Comptonization are not dominant in Sco X-1, but it is likely to be important in other sources, and students should be aware that there is a considerable literature on the subject (Shakura and Sunyaev 1973, Pringle et al. 1973).

A gas cloud with $n \gtrsim 10^{16}$ cm^{-3} would cool itself very quickly, and so there must be some continuing source of energy in Sco X-1. It was quickly suggested that this source might be a binary star, with gas escaping somehow from the surface of a more or less "normal" component and falling into the gravitational potential well of a collapsed

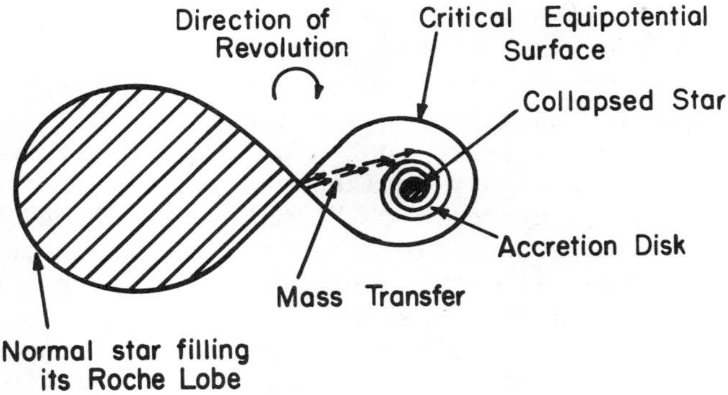

Fig. 6.6 Sketch of an X-ray binary with accretion disc.

companion, possibly a white dwarf or supernova remnant. There are several variants of this scheme; the most popular is sketched in Figure 6.6. The normal star in this close binary has swollen in the course of its evolution until it fills its Roche lobe (its side of the critical equipotential surface). Tidal torques have forced it to corotate with the revolution of the system, and matter is continually transferred at the neutral point. Studies of mass transfer and its effects upon stellar evolution in such X-ray binaries have been done recently, especially by van den Heuvel and coworkers (van den Heuvel and de Loore 1973), and the subject is taking its rightful place in the lore of binary stars.

The transferred gas, falling into the deep potential well, heats up, and its angular momentum compels it to form a swirling disk in the plane of the system. It circles the collapsed star in roughly Keplerian orbits, but viscosity causes it to drift slowly inward and be accreted. Many papers on the physics of such disks have been written (e.g. Shakura and Sunyaev 1973, Pringle et al. 1973, Page and Thorne 1974, Rees 1974), but understanding of the disk structure is not very complete, principally because the viscosity of such a medium is uncertain even to order of magnitude.

The evidence that Sco X-1 is in fact a binary is weak, despite many searches. Very recently two different binary periods have been claimed: 3.9 days (Lyuty et al. 1975) and 0.787 days (Gottlieb et al. 1975). As we shall see, however, some other X-ray sources certainly are binaries, and the accretion-disk theory has been applied extensively to these.

One way to test the basic idea of a thermal-bremsstrahlung X-ray source is to look for characteristic X-ray line emission from highly-ionized heavy atoms in the thermal plasma, e.g. the hydrogenic "Lyman-alpha" line ($h\nu = 6.97$ keV) of Fe^{+25}. The emissivities in these lines can be calculated. Unfortunately the detection of such lines in an optically thick source like Sco X-1 is fraught with difficulties (Felten et al. 1972). The narrow X-ray lines are easy prey to Comptonization, because a single Compton scattering shifts a line photon by hundreds of eV. Detection is likely to succeed first in large, diffuse, optically thin sources like the Cygnus loops (Bleeker et al. 1972, Stevens et al. 1973).

6.3. Crab Nebula

The Crab Nebula (3U 0531 + 21, Davies and Smith 1971) is known to be the remnant of a supernova observed by the Chinese in 1054 A.D. The spectrum of the Crab from radio through optical and X-ray frequencies is shown in Figure 6.7 (Woltjer 1970). This is very different from Sco X-1. At $D \simeq 2$ kpc, the luminosity is $P_{XR} \simeq 5 \times 10^{37}$ erg sec^{-1}. Note that the spectrum looks rather like two power laws of different spectral indices s, roughly 0.25 and 1.2, joined around 10^{14}Hz. In fact it looks rather like the synchrotron spectra of Figure 5.29. Now the radio and optical flux, because of its strong polarization, is believed to be synchrotron radiation from fast electrons in the remnant. It seems likely that the X-rays are also part of the synchrotron spectrum. If we set $\nu_t = 10^{14}$ Hz in equation (5.76), and use the age of the supernova remnant ($\simeq$ 900 years) for t, we find $H \simeq 2.4 \times 10^{-4}$ gauss. This is reasonably close to the mean field strength in the synchrotron-emitting regions of the Crab inferred on other grounds (Woltjer 1958, Fazio 1973). This suggests that the Crab is in fact an example of a synchrotron source with an evolved spectrum showing curvature related to the age of the source, as discussed in Section 5.9. The X-ray spectrum is too steep to fit the curve in Figure 5.29 corresponding to continuous, constant injection of fast electrons, but too flat to fit the curve for no injection subsequent to the outburst 900 years ago. With an injection rate decreasing in time, a suitable fit can be made (Tucker 1967).

This is all oversimplified, because we should consider the changes in physical conditions in the Crab (e.g. H) over 900 years, and particularly the electron energy losses due to expansion of the nebula (Kardashev 1962). But the following rough calculation is so simple as to be rather general. Consider those electrons which are producing strong synchrotron radiation at frequency ν Hz. Their energy E is given approximately by equation (5.22), with $\omega_c = 2\pi\nu$. Their rate of energy loss, $|dE/dt|$, is then given by equation (5.15). The ratio $E|dE/dt|^{-1}$ has dimensions of time and can be expressed as a function of ν and $H_\perp$:

$$t_s(\nu, H_\perp) \equiv E|dE/dt|^{-1} =$$

$$1.0 \times 10^{12} \sec[\nu(\mathrm{Hz})]^{-1/2}[H_\perp(\mathrm{gauss})]^{-3/2}. \quad (6.5)$$

This is the approximate lifetime against synchrotron energy loss for particles radiating at ν. In the Crab, with $H_\perp \simeq 2 \times 10^{-4}$ gauss, the X-ray electrons ($\nu \sim 10^{18}$Hz) have $E \sim 2 \times 10^{13}$eV and $t_s \sim 10$ years. Thus these high-energy electrons must have been produced within the nebula in recent years. This conclusion is credible, since the nebula contains a pulsar which presumably accelerates cosmic rays. In fact, changes, perhaps due to hydromagnetic waves generated by pulsar activity, are seen in the optical "wisps" near the pulsar over a few years (Scargle and Harlan 1970).

The size of the Crab as an X-ray source is relevant to this discussion. It is known that at a few keV the Crab is not a point source, but is extended over at least 1' of arc along the major axis of the nebula; along the orthogonal axis the extent is unknown but may be similar (Kellogg 1971). At $D \simeq 2$ kpc, $1' \simeq 2$ light years, so the spatial extent of the radiating electrons is a substantial fraction of the distance they are able to travel at speed c in their lifetime $t_s \sim 10$ years. Unless they are accelerated throughout the nebula, they must then stream rather effectively through the nebular field from their source (presumably the pulsar) to fill so large a volume. This is rather surpris-

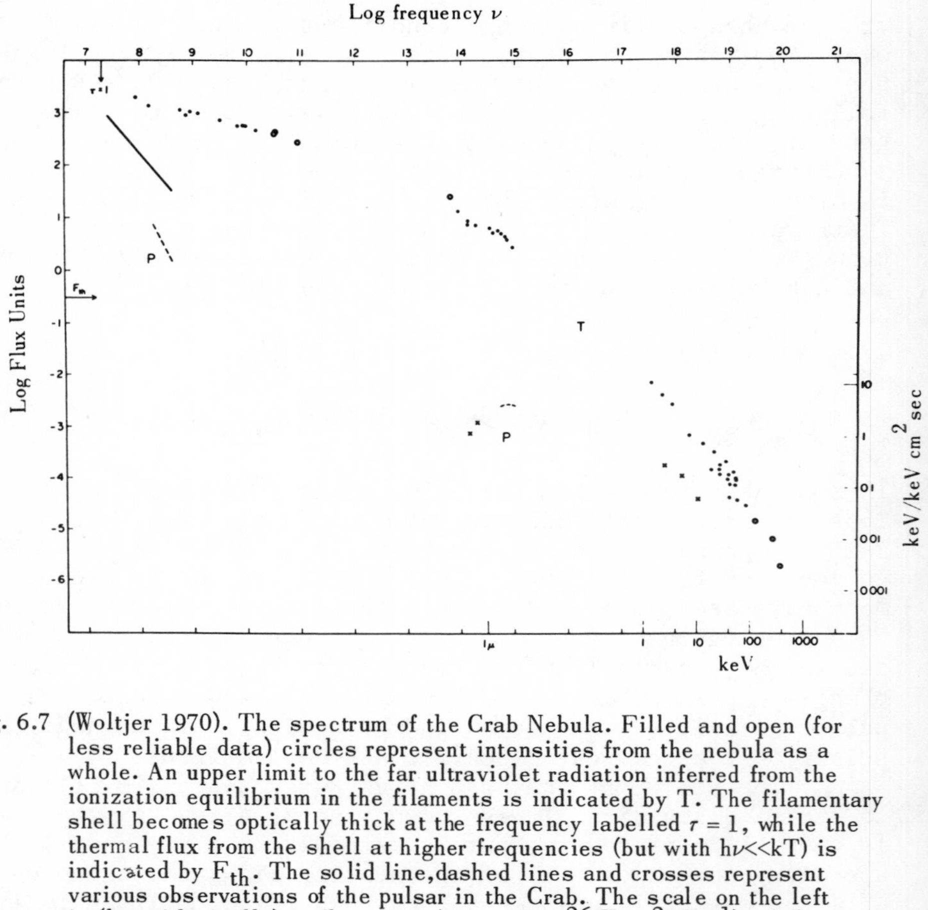

Fig. 6.7 (Woltjer 1970). The spectrum of the Crab Nebula. Filled and open (for less reliable data) circles represent intensities from the nebula as a whole. An upper limit to the far ultraviolet radiation inferred from the ionization equilibrium in the filaments is indicated by T. The filamentary shell becomes optically thick at the frequency labelled $\tau = 1$, while the thermal flux from the shell at higher frequencies (but with $h\nu \ll kT$) is indicated by F_{th}. The solid line, dashed lines and crosses represent various observations of the pulsar in the Crab. The scale on the left is (logarithmically) in flux units (1 f.u. = 10^{-26} W m^{-2} Hz^{-1}), while the scale at the top gives the $\log\nu$, with the frequency ν in Hz. For the X-ray region the intensities are also shown in keV $(cm^2$ sec keV$)^{-1}$ and the photon energies in keV. All optical and infrared data have been corrected for 1.5 mag interstellar absorption in the visual.

ing, as one might have expected a slow diffusive motion in the tangled nebular field. It will be interesting to see whether the source is smaller at higher X-ray energies (cf. equation (6.5)); observations are under way at this writing (Ricker et al. 1975, Davison et al. 1975).

Observers at Columbia University (Novick et al. 1972) have measured the linear X-ray polarization of the nebula. This is 15% ± 5% in position angle 156° ± 10°, and is consistent with the net polarization of the optical continuum from a central circular region 1' in diameter (comparable to the X-ray size). This is a powerful argument for the synchrotron character of the X radiation, since equation (5.42) predicts strong, frequency-independent polarization. (note that 15% is much smaller than the value predicted by equation (5.42); this is because the nebular H-field is complicated, while the equation applies to a uniform field). The structure and evolution of the Crab present many interesting problems (Trimble and Rees 1970, Felten 1974), which we must omit here.

We should mention briefly that interstellar X-ray absorption has been detected in the spectrum of the Crab. Since D is fairly well known, and since 21-cm absorption measurements on the radio source yield an independent determination of the column density N_H of neutral hydrogen, this object is a useful check on theories of the X-ray absorption. The data raise some difficulties in their interpretation; one possible resolution is discussed by Margon (1974).

6.4. Hercules X-1

Hercules X-1 (HZ Her, 3U 1653 + 35) is a true X-ray binary (Tananbaum et al. 1972, Giacconi et al. 1973). The X-rays come in pulses with a 1.24-second period. Every 1.7 days, these pulses disappear quite abruptly (within <12 min) and do not reappear for about 6 hours. These disappearances can be seen in Figure 6.8. The X-ray pulse period can be measured accurately, and it shows a regular variation back and forth; this cycle also takes 1.7 days. The obvious interpretation is that an emitter of precisely timed X-ray pulses - - an "X-ray pulsar" - - is in orbit around another body and is being eclipsed by it for 6 hours in every 1.7-day period. The cycle of change in the pulse period is interpreted as Doppler shift due to the orbital motion. The

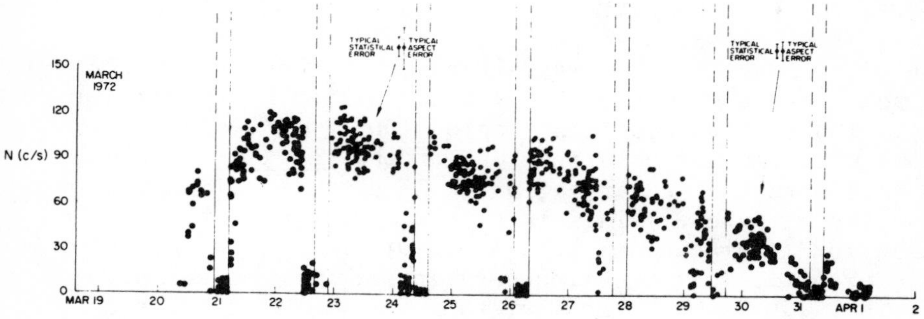

Fig. 6.8 (Giacconi et al. 1973). Her X-1 intensity (counts/sec) data during one ON cycle. The vertical lines represent the orbital eclipses, whose positions are determined accurately from the pulsation Doppler analysis. Typical errors appropriate for different groups of data are shown; the statistical error bar is relevant for point-to-point comparisons, and the aspect error bar for day-to-day comparisons. It should be noted that most intensity points below about 10 counts/sec are upper limits.

amplitude of the Doppler shift gives the radial component of the X-ray object's orbital velocity as 170 km sec^{-1}, and the sinusoidal shape of the Doppler variation tells us that the orbits are nearly circular rather than highly elliptical. Newton's laws, in the form of the classical theory of spectroscopic binaries, can be applied to these data, and yield a value for the following quantity, called the mass function:

$$\frac{M_{opt}^3 \sin^3 i}{(M_X + M_{opt})^2} = 0.85 M_\odot \ . \qquad (6.6)$$

Here "X" denotes the X-ray object and "opt" its putative "optical" or normal companion. i is the inclination of the axis of the binary orbit to our line of sight; it must be fairly close to 90° for eclipses to occur. One also concludes that the radial component of the X-ray object's orbital radius is $r_X \sin i \simeq 4 \times 10^{11}$ cm. Note that all this information is derived from the X-ray observations alone.

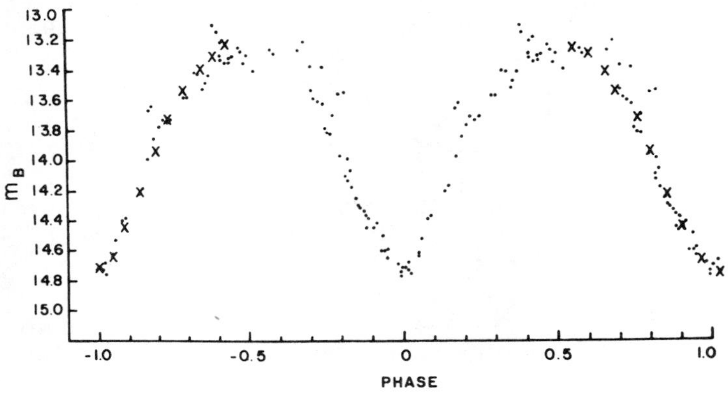

Fig. 6.9 (Bahcall and Bahcall 1972). Light curve for HZ Her. The dots are measured magnitudes; the crosses are calculated from a theoretical model. Measurements over six weeks have been plotted modulo the X-ray period of 1.70017 days. Individual measurements are plotted twice in order to show two complete cycles. The zero in phase is the same as the time of maximum occultation of Her X-1 reported by Tananbaum et al. (1972).

This source is identified with the optical variable HZ Herculis, which varies by about 1.5 mag with a period of 1.7 days. One set of observations is shown in Figure 6.9 (Bahcall and Bahcall 1972). The optical minima, which are rather sharp, occur at the midpoints of the X-ray eclipses. This suggests a binary model like that of Figure 6.6, in which the collapsed star (the "pulsing" X-ray source), moving around the normal star, produces anomalous heating of the normal star on the side turned toward the X-ray source. This hot "cap" follows it around, so that when the X-ray object is on the near side, the hot cap is turned toward us, and we see maximum light. Arguments based on this model lead to the conclusion that the normal star is a late A or early F type at a distance of several kpc (which implies a 2-to-10-keV X-ray luminosity $P_{XR} \sim 2 \times 10^{36}$ erg sec^{-1}), with a mass $M_{opt} \simeq 2.3 M_{\odot}$. Then $M_X \simeq 1.5\ M_{\odot}$. This could be checked if the Doppler radial-velocity curve for the <u>normal star</u> could be obtained, but its (optical) spectral lines vary in shape and it has not been possible to measure the radial velocities accurately. The X-ray object may well be a neutron star, since stable neutron stars are believed to exist in the mass range $< 2\ M_{\odot}$.

There are many complications in the behavior of this source which are not understood theoretically. The shape of the optical light curve, particularly around the minima, is not well understood. Other difficulties arise from the complicated X-ray behavior (Figure 6.8). There is a 35-day cycle in which the X-ray pulses appear and disappear at their expected times for 12 days, then disappear entirely for 23 days. There are also irregular "dips"--sometimes the source _is_ seen when not expected! There are several theories of the 35-day modulation; one popular theory is that the X-rays are emitted in a beam for some reason, rotating once in 1.24 sec., and the beam is precessing (Strittmatter _et al_. 1973), so that sometimes it sweeps over us and sometimes not. We shall make a few more remarks about theories of Her X-1 in Section 6.6. The course Cen X-3 (3U 1118 -60) is similar in many ways to Her X-1, but we must omit this. These are the only "pulsing" X-ray binaries known.

6.5. Cygnus X-1

Two recent papers on the distance to Cygnus X-1 (HDE 226868, 3U 1956+35) (Margon _et al_. 1973, Bregman _et al_.1973) give numerous references to the literature on it. The optical object in this case is apparently a more or less normal 9th-magnitude B0 supergiant at a distance $\gtrsim$ 2.5 kpc, with mass $\sim 30 M_\odot$. The X-ray source, which then has $P_{XR} \gtrsim 10^{37}$ erg $\sec^{-1}$, does not pulse periodically and does not eclipse. Nevertheless it is thought to be an X-ray binary, because the star is an optical spectroscopic binary with a period of 5.6 days. Once again the classical theory of binaries can be applied; this time there is no Doppler information on the X-ray source, because it has no pulses or X-ray lines to serve as reference frequencies, but there _is_ optical Doppler information on the star. The positions of M_X and M_{opt} are then reversed from what they were in equation (6.6), and one finds for the mass function

$$\frac{M_X^3 \sin^3 i}{(M_X + M_{opt})^2} = 0.23 M_\odot . \qquad (6.7)$$

We see that if M_{opt} is large, M_X must also be quite large. If $M_{opt} \sim 30M_\odot$, we have $M_X \sim 6M_\odot$. This is too massive to be a neutron star, and the common belief is that Cyg X-1 is probably a black hole in orbit around a supergiant! The X-ray source must be compact, because there are irregular X-ray variations in times as short as 50 msec, which imply that the source cannot be larger than 10^9 cm -- about the size of a white dwarf. There is in fact an alternative theory that M_X is a "spinar" -- a massive white dwarf, too massive to be stable if static, which is stabilized by rapid rotation (Brecher and Morrison 1973). Even faster X-ray variations have been reported, and if confirmed might favor the black hole. A third view is that Cyg X-1 is a triple system (Bahcall et al. 1974). In this case the collapsed object need not have a large mass and can be a neutron star. This is comforting to persons who dislike the idea of a black hole!

6.6. Binary X-Ray Sources

The objects 3U 1700-37, 3U 0900-40, and 3U 0115-73 (SMC X-1, in the Small Magellanic Cloud) are all more or less similar to Cyg X-1. In addition Cyg X-3 (3U 2030+40) is a possible X-ray binary. Thus there are six or seven known X-ray binaries in all. These, together with Sco X-1 and Cyg X-2 (3U 2142+38), comprise the "compact X-ray sources", all of which are thought likely to be binaries. There are two useful reviews (Rees 1974, Blumenthal and Tucker 1974) which give further information on theories and observations of these sources.

It may be useful if we summarize here the general theoretical outlook on these sources as a group. There are two broad classes of theories. The first view, currently more favored, is that all the compact sources are variations on the theme of an accretion disk around a collapsed object in a binary system. The X-rays are viewed as thermal emission from the hot accreting gas. The spectrum need not be that of optically thin thermal bremsstrahlung, because T is expected to vary over the disk, with the inner part hottest. Comptonization and other optical-thickness effects are also important.

The pulsing sources (Her X-1 and Cen X-3) are viewed as neutron stars (Lamb et al. 1973; Figure 6.10). Sufficiently near the neutron star, the strong magnetic field

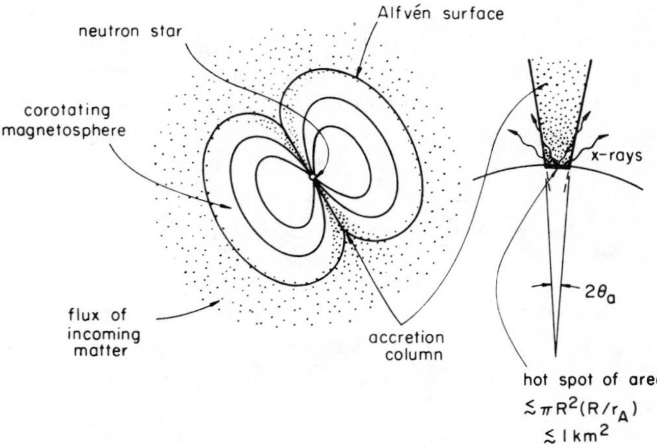

Fig. 6.10 (Lamb et al. 1973). Sketch depicting some of the basic features of accretion onto a magnetic neutron star. The left side of the figure is an overall view of the magnetosphere showing the location of the Alfven surface, the cusplike configuration of the magnetic field above the magnetic poles, the flow of accreting matter (dots) across the Alfven surface, and the formation of accretion columns above the magnetic poles. To the right is a close-up view of an accretion column near the stellar surface. The accreting matter collides with the surface of the neutron star over an area $< \pi R^2(R/r_A)$, where R and r_A are, respectively the radius of the star and the radius of the Alfven surface. The result is a hot spot which leads to anisotropic X-ray emission.

(10^{10}-10^{12} gauss) imbedded in the neutron star must control the flow of the accreting matter. The disk therefore terminates at some inner surface (the "Alfven surface"), and from here on in the accreting gas is channeled along field lines onto the magnetic poles of the neutron star, forming "accretion columns" (Figure 6.10). X-rays are then emitted strongly at the poles.

The magnetic field is imbedded in the spinning neutron star and must corotate with it. This may well cause the X-ray beam to rotate, sweeping past us and producing the pulses - - provided the X-rays can be generated in a beam pattern at all. There are various suggestions of how this happens. Optical-thickness effects in the accretion column might cause the X-rays to escape preferentially at the sides of the column rather than along its length, thus producing a fan beam. On the other hand, the magnetic field, if strong enough, may reduce the electron-scattering cross section for photons travelling along the column (parallel to the field) below the Thomson value. (We can understand this classically because, when ω_G in equation (5.13) is higher than the radiation frequency, the electrons are not free to move linearly like dipole scatterers in responding to radiation travelling along the field). A pencil beam might result. These things have not been worked out in any detail, partly because theorists are not sure whether a pencil or a fan beam would be more useful in accounting for the observations!

The non-pulsing sources are viewed as black holes. The black hole has no attached magnetic structure; material at the inner edge of the disk just disappears into the hole. Most of the X-rays, particularly the hard X-rays, must be produced in this inner hot part of the disk, where general-relativistic effects of the black hole are large. Obviously X-ray binaries are of great interest if their radiation can tell us about the existence and properties of black holes.

The other, contrasting view of binary sources is that they are pulsating white dwarfs in binary systems, experiencing accretion, possibly quasi-spherical accretion. The accreting matter somehow triggers instabilities in the star, and it pulsates (irregularly, or, in the case of the pulsing sources, regularly!), generating shock waves which heat its accreting "atmosphere" and cause X-ray emission. In the cases where the derived M_X is too large ($> 1.2M_\odot$, the Chandrasekhar limit) to be a classical white dwarf, it is suggested that the dwarf is stabilized by the centrifugal force of its rapid rotation (Brecher and Morrison 1973). Then perhaps the rotation can furnish the "clock" for the X-ray pulses, and no regular pulsation is needed! We omit further details of these models, which are in large part very sketchy anyway.

Since all these ideas are based on accretion, we shall present one simple calculation (Cameron and Mock 1967) which suggests that the accretion approach is correct. Consider spherically symmetric gravitational accretion onto a body of mass M_X, and suppose that this body radiates power P_{XR} isotropically. The radiation exerts outward pressure on the infalling matter; if this were to equal the gravitational pull, there would be no net force to maintain the accretion. If the gas is ionized hydrogen and the main opacity is electron scattering, it is given by the Thomson cross section σ_T. Equating the two forces, per atomic mass of H at a distance r, gives

$$\frac{P_{XR}}{4\pi r^2 c} \sigma_T = \frac{GM_X m_H}{r^2} \quad . \tag{6.8}$$

We conclude that the power of an accretion source should be self-limiting:

$$P_{XR} \lesssim \frac{4\pi G M_X m_H c}{\sigma_T} = 1.25 \times 10^{38} \left(\frac{M_X}{M_\odot} \right) \text{ erg sec}^{-1} \quad . \tag{6.9}$$

This is the "Eddington limit" (Eddington 1926). It is comforting that most of the X-ray binaries lie one or two orders of magnitude below this limit. Only in the case of the most luminous source, SMC X-1, is there a marginal discrepancy (Osmer and Hiltner 1974). There may be ways to violate this limit by a modest factor in the asymmetric models discussed above. Note also that the Eddington limit represents a static balance and could certainly be violated locally in dynamic flow patterns.

6.7. Other Galactic Sources; Gamma-Ray Sources

There are quite a few other individual galactic sources of interest, but we shall omit these, along with the interesting subject of the general distribution (in luminosity and position) of the galactic X-ray sources (Seward et al. 1972, Margon and Ostriker 1973). There seems to be a "ridge" of unresolved X-ray emission along the galactic plane, possibly due to a population of weak unresolved sources, but the situation with regard to this is somewhat confused at present (Bleach et al. 1972, Silk 1973). The total strength of X-ray sources in our Galaxy, integrated over all observed $h\nu$, seems to be about 5×10^{39} erg sec^{-1}.

The ridge of the Galaxy has definitely been detected in the gamma-ray band ($h\nu \gtrsim 10$ MeV). These gamma rays are probably produced as secondaries in collisions of cosmic-ray particles with atoms of gas in the disk, and are therefore of special interest to cosmic-ray physicists (Stecker and Trombka 1973, Bignami and Fichtel 1974).

Although gamma-ray astronomy is still in its infancy, a word about other known gamma-ray sources is necessary. The Crab Nebula has been detected over a wide range of gamma-ray energies (McBreen _et al._ 1973). There is a steady component which appears to be an extension of the synchrotron X-ray spectrum, and also a pulsed component produced by the pulsar.

Perhaps the most interesting gamma-ray sources are the recently discovered transient or "burst" sources (Strong _et al._ 1974). The typical duration of these burst sources is of order one second. More than 30 have been observed since 1967. They have been seen below 10 keV as well as above 1 meV (Metzger _et al._ 1974), and most of the energy seems to lie between 100 keV and 1 MeV, so that they should be thought of as hard X-ray bursts. The locations of some on the sky are known, but only roughly, and no conclusive identifications have been made with any celestial objects. They are not concentrated strongly to the galactic plane, and so we suspect that they are either extragalactic (flashes from distant supernovae?) or else galactic objects within a few kpc at most. Theories of these bursts are proliferating; the accumulation of additional events should shed more light on them.

6.8. Extragalactic Sources

Many extragalactic X-ray sources have been identified (Kellogg 1973); in general these are either clusters of galaxies or strong radio sources. The radio galaxies Virgo A (M87) and Centaurus A (NGC 5128) are examples of the latter. The X-ray power of the M87 source is $\sim 2 \times 10^{42}$ erg sec^{-1}, and its angular size is less than or equal to that of the optical galaxy. It may be coextensive with the famous "jet" which is seen in radio and optical observations near the nucleus of this galaxy. The X-ray flux lies roughly on an extrapolation of the radio and optical synchrotron spectrum, and it is likely (though not certain) that here we have another synchrotron X-ray source like the Crab but on a much larger scale. This raises time-scale and particle-diffusion problems rather more acute than those in the Crab (Felten 1970, Turland 1975). Besides the small X-ray source associated with the galaxy, there is a large one, half a degree across, with P_{XR}~5×10^{42} erg sec $^{-1}$, which may be associated with the Virgo cluster of galaxies as a whole (Catura et al. 1974).

Centaurus A (Lampton et al. 1972) has $P_{XR} \sim 6 \times 10^{41}$ erg sec^{-1}. The X-ray source lies on or near the optical galaxy, not on the north lobe of the associated large double radio source as once thought. The X-ray flux in this case lies well below the extrapolation of the synchrotron radio spectrum, but the X-ray spectrum does seem to have a power-law shape.

It is possible that here we are seeing inverse Compton radiation from the same power-law electrons that generate the radio spectrum of Cen A. Recall equation (5.11) and the statement there that

$$\varepsilon \sim \gamma^2 \varepsilon_0 , \qquad (6.10)$$

where ε_0 is the energy of the original (ambient) photon and ε that of the Compton-scattered photon. Equations (5.11) and (6.10) for Compton scattering are isomorphic to equations (5.15) and (5.22) for synchrotron radiation. Essentially for this reason, a power-law electron distribution, which generates a power-law synchrotron spectrum in the optically thin case (Section 5.4), also generates a power-law inverse Compton spectrum with the same slope (Felten and Morrison 1966). (A good review of the physics of the synchrotron and Compton processes is given by Blumenthal and Gould [1970].) Any photons which are present in the region of the source can serve as ambient photons ε_0 and contribute to u_{rad} in equation (5.11). In particular, the cosmic 2.7°K radiation contributes to u_{rad}. It is possible that in Cen A we are seeing X-rays produced in Compton collisions between fast electrons and cosmic black-body photons. The number of fast electrons in the radio source is not well known, and it must be very large if this suggestion is correct.

Several other well-known extragalactic objects are X-ray sources. The Seyfert galaxy NGC 4151 has P_{XR}~1×10^{42} erg sec $^{-1}$. The quasar 3C 273 is also a source, and if its distance is as given by its redshift, then $P_{XR} \sim 10^{46}$ erg sec $^{-1}$! M31, the spiral galaxy in Andromeda, has $P_{XR} \sim 3 \times 10^{39}$ erg sec^{-1} and is apparently much like our Galaxy as an X-ray emitter. The Large and Small Magellanic Clouds contain several sources, one of which has already been mentioned. For most extragalactic sources, the spectra are not yet well enough known to permit more than speculation about the nature of the emission (Kellogg 1973, Laros et al. 1973).

There is a large family of sources which appear to be clusters of galaxies (Kellogg et al. 1973, Fabian et al. 1974). We have already mentioned the Virgo Cluster. Clusters in Centaurus, Coma Berenices and Perseus and the cluster Abell 2256 appear to be similar, and there are about 15 others which are probable or possible analogs. There are two leading theories of the cluster sources. It is possible (Silk 1973) that these clusters contain large amounts of hot gas at $T \sim (3\text{-}10) \times 10^7\,°K$, possibly heated over cosmic time by the galaxies plowing back and forth within the clusters (Felten et al. 1966). This gas could generate thermal-bremsstrahlung X-rays. Brecher and Burbidge (1972) have proposed that instead these clusters, which are known to be weak extended radio sources, contain large numbers of fast electrons, perhaps escaped from active galaxies within the clusters, which occupy the cluster volume and produce the X-rays by Compton interactions with the 2.7 °K photons. As in the case of Cen A discussed above, this idea requires that the number of these electrons be very large (Harris and Romanishin 1974). Evidence seems about equally divided between these two theories at present.

6.9. The Diffuse Background

The diffuse (unresolved) flux was discovered quite early. A recent compilation of data on its spectrum (Silk 1973) is shown in Figure 6.11. The spectrum appeared at first to be a power law with slope about -1.2. Now there is a good deal of structure, although the data scatter rather widely and some of the structure may be illusory. Note in particular the apparent "bump" or change in slope around 30 keV. A critical review of the data (Schwartz and Gursky 1973) seems to confirm the reality of this. The flux above 1 keV is isotropic as nearly as has been determined (Fabian and Sanford 1971), which implies that it is extragalactic and probably arises at great distances. It is therefore of cosmological interest.

The observed strength of the background flux is $F \simeq 40$ keV $(cm^2\ sec\ sr)^{-1}$ over 2-10 keV. This corresponds to an energy density $u \simeq 2 \times 10^{-5}$ eV cm^{-3}. Over the whole X-ray

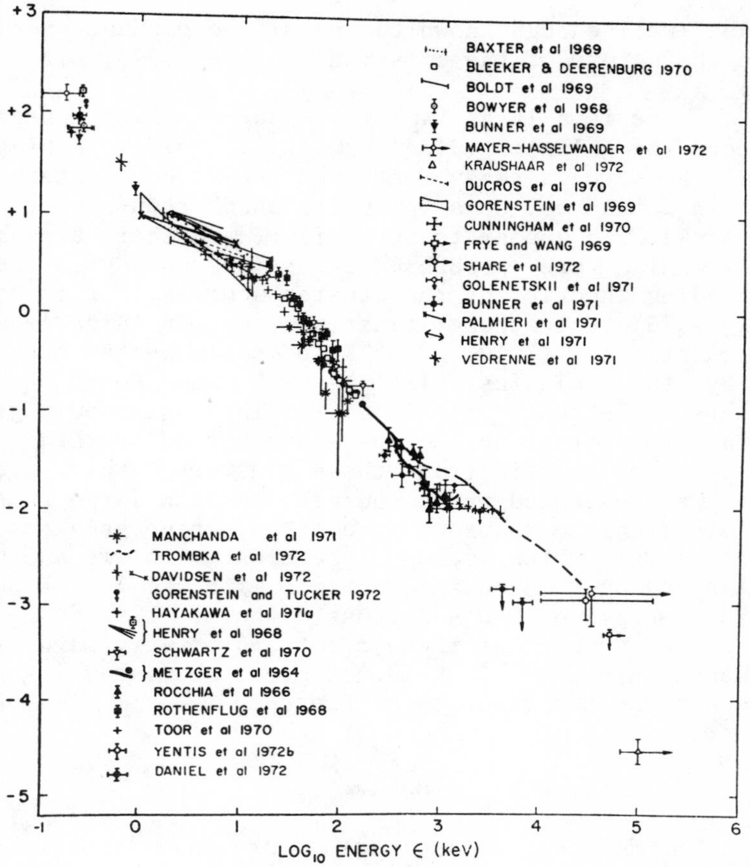

Fig. 6.11 (Silk 1973). Observations of the isotropic X-ray background. The high-galactic-latitude fluxes near 0.25 keV are uncorrected for interstellar absorption.

spectrum it would be ~ 10^{-4} eV cm^{-3}. This is surprisingly large. To see why, we make a simple calculation. Suppose that we have some X-ray sources distributed over intergalactic space, the mean emitted power per unit volume being J erg $(cm^3\ sec\ sr)^{-1}$, and suppose that these sources extend out to some large distance R. Then the flux we receive is F = JR erg $(cm^2\ sec\ sr)^{-1}$, if absorption and red shift can be neglected. (It is easily shown that X-ray absorption at intergalactic densities must be negligible out to a Hubble distance at $h\nu$ > several keV). Now for sources at cosmological distances, an effective (approximate) cutoff is provided by red shift at a distance $R \sim (1/2) c H_0^{-1} \equiv 1/2\, R_H$. If the number density in space of some sources (e.g. normal galaxies) is n, and their mean luminosity is P_{XR}, then $J = (4\pi)^{-1} n P_{XR}$, and the expected flux due to unresolved distant sources of this kind is

$$F \sim \frac{n P_{XR} R_H}{8\pi} . \tag{6.11}$$

For normal galaxies something like our own, if $H_0 = 75$ km sec^{-1} Mpc^{-1}, we have $n \sim 3 \times 10^{-2}$ Mpc^{-3} (Felten 1973), and $P_{XR} \simeq 3 \times 10^{39}$ erg sec^{-1} in 2-10 keV. (M31 is about the same.) Then we find $F \sim 0.9$ keV $(cm^2\ sec\ sr)^{-1}$. Comparing this with the observed F given above, we see that distant normal galaxies could not provide more than a few per cent of the observed background.

This rough comparison has been made for several classes of objects, to test the viability of these "superposition" theories of the X-ray background (Schwartz and Gursky 1973). The conclusion is that some 20 per cent of the 2-10 keV background can be accounted for in this way, mostly by Seyfert galaxies and rich clusters of galaxies. One might even argue that 20 per cent equals 100 per cent! Unfortunately, it is even more difficult to account for the background at $h\nu$ > 10 keV in this way; the extragalactic discrete sources in general do not seem to be as hard as the background spectrum, though the Seyfert galaxy NGC 4151 is indeed reported as a hard source (Baity *et al.* 1974). It seems likely that there is an additional large contribution to the background, arising in discrete sources of a type not yet observed, or possibly not in discrete sources at all.

One example of the former type of theory is the supernova theory. If supernovae occur in our Galaxy every 40 years roughly, and if a supernova were to produce $\sim 2 \times 10^{50}$ erg in a rather hard X-ray spectrum, then the *time-average* luminosity of the Galaxy due to these outbursts would be $\sim 2 \times 10^{41}$ erg sec^{-1}, about 40 times its steady P_{XR} which we observe! The background could then be made up of supernovae flashing in millions of distant galaxies (Tucker 1970). This theory can be tested by X-ray observations of supernovae in neighboring galaxies, and eventually in our own.

Another superposition theory, for the hard X-ray ($h\nu \gtrsim 1$ MeV) and gamma-ray background, was proposed by Sunyaev (1970). He attributed strong gamma-ray bremsstrahlung emission to the nuclei of Seyfert galaxies, speculating that they might contain thermal plasma with $kT \sim 20$ MeV. Another process must produce the softer X-rays.

There is a wide variety of theories attributing the background to processes occurring generally in the intergalactic medium, or in some diffuse regions (not within galaxies). Other theories attribute it to galaxies (or other condensed objects) in the different physical conditions prevailing at large z (earlier in the history of the universe). These theories are all quite speculative and do not deserve lengthy discussion in a short introduction to the subject, but we shall catalog them briefly. The inverse-Compton theory, in several forms (Felten and Morrison 1966, Brecher and Morrison 1969), imports that fast intergalactic electrons produce the background by scattering 2.7 °K photons. The electrons have presumably leaked out from sources in radio galaxies or normal galaxies. This theory applied to the background, just as when applied to Cen A or to the cluster sources, presses its advocates rather hard to provide enough fast electrons. Another variant of the inverse-Compton theory (Felten and Rees 1969) attributes the background to radio galaxies at $z \simeq 3$-5. These theories are all embarrassed by the apparent "break" in the observed spectrum at 30 keV, as there is no natural way to produce it (Felten 1973).

Silk (1970) has suggested that the background is <u>nonthermal bremsstrahlung</u> from fast electrons in the intergalactic medium. It helps in accounting for the 30-keV break if the nonthermal spectrum of electrons is assumed to be present at an early time ($z \sim 10$). These numerous electrons, besides emitting bremsstrahlung X-rays, must heat the intergalactic gas strongly. The resulting severe difficulties for the theory are discussed by Setti and Rees (1970).

Of several theories relating the hard X-ray and gamma-ray background to cosmic rays in the early universe, the best known is that of Stecker (1969). Cosmic-ray protons collide with gas protons and produce (<u>inter alia</u>) π^0-mesons, which decay into gamma rays. These gamma rays have a spectrum peaked around 70 MeV, but if the reactions occur at $z \sim 100$, the gamma rays are red-shifted to $h\nu \sim 0.7$MeV at present, and a good fit to the observed gamma-ray spectrum can be obtained. The theory requires a very large cosmic-ray flux at $z \sim 100$, but like anything else at that epoch this is hard to deny. Another process must produce the X-rays at $h\nu << 1$ MeV.

A related theory (Stecker et al. 1971, Stecker 1973) relies upon matter-antimatter annihilation to produce the gamma rays. Once again, large contributions must be received from $z \sim 100$. The theory is of course completely dependent upon the existence of large amounts of antimatter in the early universe.

One rather simple theory of the X-ray "break" at 30 keV deserves special mention, namely that of a very hot intergalactic medium. In general-relativistic cosmology, the present mean mass density in space required to make the universe closed rather than open is

$$\rho_{c\ell} = \frac{3H_0^2}{8\pi G} , \tag{6.12}$$

the "closure density". For the recent value of $H_0 \simeq 50$ km sec^{-1} Mpc^{-1}, we have $\rho_{c\ell} \simeq 4.7 \times 10^{-30}$ g cm^{-3}, which implies a particle density $n \sim 3 \times 10^{-6}$ cm^{-3}. If $H_0 = 50$ or a little less, and the density of intergalactic gas is $\rho = \rho_{c\ell}$, and its temperature is $T \simeq 3 \times 10^8$ °K, the thermal -bremsstrahlung X-ray flux received from the intergalactic medium will be roughly what is observed at 30 keV (Cowsik and Kobetich 1972). In making this calculation it is necessary to be a little bit careful about the cosmological integration and about the thermal history of the intergalactic gas (Field 1972). The advantage of this theory is that the exp $(-h\nu/kT)$ shape of thermal bremsstrahlung can give the rather sharp knee at 30 keV apparent in the data, provided $T \simeq 3 \times 10^8$ °K. One disadvantage is that other sources must then be sought for most of the X-rays well below and well above 30 keV. Another is the difficulty of accounting for the large heat content implied.

6.10. Soft X-Ray Background

Finally I must say something about the X-ray background at $h\nu < 1$ keV. The earliest observations showed that the background at these energies is not isotropic. This was expected, for absorption by gas in the galactic disk is not negligible (see the discussion in Section 6.2). If the background outside the Galaxy were isotropic, we would expect to see the maximum flux penetrating to us from the directions of the galactic poles (where the column density of absorbing gas is lowest), and much lower fluxes coming from directions near the plane. For a time it appeared (Felten 1973) that the observations were reasonably in accord with this picture. It was thought that the isotropic soft X-rays incident upon the Galaxy might be thermal bremsstrahlung from a hot intergalactic medium with $\rho \sim \rho_c$ and $T \sim (1\text{-}4) \times 10^6$ °K (kT ~ few hundred eV). It is still possible that such a medium, or some other extragalactic source, does contribute substantially to the soft X-rays seen in the polar directions - - though, of course, $T \sim 10^6$ °K is preempted if we believe with Cowsik and Kobetich that $T \sim 10^8$ °K. Recent work by the Wisconsin group (Williamson _et al_. 1974) has, however, shown that the angular structure of the soft X - rays is exceedingly complex, and suggests that much if not all of the flux originates in the galactic disk and has nothing to do with cosmology. One possibility is that large regions of the _interstellar_ medium are at elevated temperatures (~ 10^6 °K) and produce _galactic_ thermal bremsstrahlung. Observations of stellar ultraviolet absorption lines by the Copernicus satellite appear to support this view.

We must expect that many more of our theoretical ideas about X-ray sources will similarly be reversed as observations accumulate in this rapidly changing field.

Bibliography and References

(The following references will be found useful, some of which are referred to in the text.)

1. Radiative Transfer and Stellar Atmospheres

Chandrasekhar, S. 1960, *Radiative Transfer*, Dover.

Fowler, R. H. 1955, *Statistical Mechanics*, Cambridge University Press, 2nd Edition pp. 659-660.

ter Haar, D. 1974, *Elements of Statistical Mechanics*, Pergamon Press.

Jefferies, J. T. 1968, *Spectral Line Formation*, Blaisdell.

Leighton, R. B. 1959, *Principles of Modern Physics*, McGraw-Hill.

Krook, M. 1955, *Astrophysical Journal 122, 488.*

Mihalas, D. 1970, *Stellar Atmospheres*, Freeman.

Swihart, T. L. 1970, *Basic Physics of Stellar Atmospheres*, Pachart.

Swihart, T. L., and Brown, D. R. 1967, *Annales d'Astrophysique 30, 659.*

Wooley, R. v.d. R. and Stibbs, D. W. N. 1953, *The Outer Layers of a Star*, Clarendon Press.

2. Stellar Interiors

Aller, L. H. and McLaughlin, D. B. (eds.) 1965, *Stellar Structure*, University of Chicago Press.

Chandrasekhar, S. 1957, *An Introduction to the Study of Stellar Structure*, Dover.

Chiu, H. Y. 1965, *Neutrino Astrophysics*, Gordon and Breach.

__________. 1968, *Stellar Physics*, Blaisdell.

Chiu, H. Y. and Muriel, A. (eds.) 1972, *Stellar Evolution*, MIT Press.

Clayton, D. D. 1968, *Principles of Stellar Evolution and Nucleosynthesis*, McGraw-Hill.

Cox, J. P. and Giuli, R. T. 1968, *Principles of Stellar Structure* (2 vols.), Gordon and Breach.

Flügge, S. (ed.) 1958, *Handbuch der Physik*, LI, Springer-Verlag.

Frank-Kamenetskii, D. A. 1962, *Physical Processes in Stellar Interiors*, NSF.

Gratton, L. (ed.) 1963, *Star Evolution*, Academic Press.

Menzel, D. H., Bhatnagar, P. L., and Sen, H. K. 1963, *Stellar Interiors*, Wiley.

Schwarzschild, M. 1958, *Structure and Evolution of the Stars*, Princeton University Press.

Stein, R. F. and Cameron, A. G. W. (eds.) 1966, *Stellar Evolution*, Plenum Press.

Swihart, T. L. 1972, *Physics of Stellar Interiors*, Pachart.

3. Gaseous Nebulae

Brocklehurst, M. 1971, *M.N.R.A.S.*, 153, 471.

Harman, R.J., and Seaton, M. J. 1966, *M.N.R.A.S.*, 132, 15.

Mihalas, D. 1972, *Non-LTE Model Atmospheres for B and O Stars* (NCAR-TN/STR-76).

Milne, E. A. 1924, *Phil. Mag.*, 47, 209.

Minkowski, R. 1968, *IAU Symposium No. 34: Planetary Nebulae*, ed. D. E. Osterbrock and C. R. O'Dell (Dordrecht: D. Reidel Publishing Co.), p. 290.

Osterbrock, D. E. 1955, *Astrophysical Journal*, 122, 235.
__________. 1974, *Astrophysics of Gaseous Nebulae*, W. H. Freeman and Co.
Panagia, N. 1973, *Astronomical Journal*, 78, 929.
Rosseland, S. 1936, *Theoretical Astrophysics*, Clarendon Press, Chapter 22.
Seaton, M. J. 1960, *Repts. Progress Phys.*, 23, 313.
Strömgren, B. 1939, *Astrophysical Journal*, 89, 526.
__________. 1948, *Astrophysical Journal*, 108, 242.
Zanstra, H. 1931, *Publ. Dom. Ap. Obs.*, 4, 209.

4. A Brief Introduction to Relativity

Adler, R., Bazin, J., and Schiffer, M. 1965, *Introduction to General Relativity*, McGraw-Hill.
Jackson, J. D. 1962, *Classical Electrodynamics*, J. Wiley & Sons.
Landau, L. D., and Lifshitz, E. M. 1962, *The Classical Theory of Fields*, 2nd edition, Addison-Wesley.
Panofsky, W. K. H., and Phillips, M. 1962, *Classical Electricity and Magnetism*, 2nd edition, Addison-Wesley.
Rindler, W. 1960, *Special Relativity*, Oliver and Boyd.
Rindler, W. 1969, *Essential Relativity*, Van Nostrand Reinhold.
Schaefer, C, 1950, *Einführung in die theoretische Physik*, Vol. III, 2nd edition, Walter de Gruyter.
Schwartz, H. M. 1968, *Introduction to Special Relativity*, McGraw-Hill.
Synge, J. L., and Schild, A. 1961, *Tensor Calculus*, University of Toronto.

5. Synchrotron Spectra

Ginzburg, V. L. and Syrovatskii, S. I. 1965, *Ann. Rev. Astron. Astrophys*. 3, 297-350.
__________. 1969, *Ann. Rev. Astron. Astrophys*. 7, 375-420.
Lequeux, J. 1967, in: *High Energy Astrophysics*, ed. DeWitt, Schatzman and Veron, Gordon and Breach.
Pacholczyk, A. G. 1977, *Radio Galaxies: Radiation Transfer, Dynamics, Stability and Evolution of a Synchrotron Plasmon,* Pergamon Press.

Pacholczyk, A. G. 1970, *Radio Astrophysics: Nonthermal Processes in Galactic and Extragalactic Sources*, Freeman.

________. 1974, in: *Planets, Stars and Nebulae*, ed. Gehrels, University of Arizona Press.

Scheuer, P. A. G. 1967, in: *Plasma Astrophysics*, ed. Sturrock, Academic Press.

Scheuer, P. A. G., and Williams, P. J. S. 1968, *Ann. Rev. Astron. Astrophys.* 6, 321-350.

6. Physics and Astronomy of Celestial X-Ray Sources

Bahcall, J. N., and Bahcall, N. A. 1972, *Astrophysical Journal (Lett.)* 178, L1.

Bahcall, J. N., Dyson, F. J., Katz, J. I., and Paczynski, B. 1974, *Astrophysical Journal (Lett.)* 189, L17.

Baity, W. A., Wheaton, W. A., and Peterson, L. E. 1974, *Astrophysical Journal (Lett.)* 199, L5.

Bignami, G. F., and Fichtel, C. E. 1974, *Astrophysical Journal (Lett.)*189, L65.

Bleach, R. D., Boldt, E. A., Holt, S. S., Schwartz, D. A., and Serlemitsos, P. J. 1972, *Astrophysical Journal (Lett.)* 174, L101.

Bleeker, J. A. M., Deerenberg, A. J. M., Yamashita, K., Hayakawa, S., and Tanaka, Y. 1972, *Astrophysical Journal* 178, 377.

Blumenthal, G. R., and Gould, R. J. 1970, *Rev. Mod. Phys*. 42, 237.

Blumenthal, G. R., and Tucker, W. H. 1974, *Ann. Rev. Astr. Ap.* 12, 23.

Bradt, H., and Giacconi, R., eds. 1973, *X- and Gamma-Ray Astronomy* (IAU Symp. No. 55) Reidel.

Brecher, K., and Burbidge, G. R. 1972, *Nature* 237, 440.

Brecher, K., and Morrison, P. 1969, *Phys. Rev. Lett.* 23, 802.

________. 1973, *Astrophysical Journal (Lett.)* 180, L107.

Bregman, J., Butler, D., Kemper, E., Koski, A., Kraft, R. P., and Stone, R. P. S. 1973, *Astrophysical Journal (Lett.)* 185, L117.

Brown, R. L., and Gould, R. J. 1970, *Phys. Rev.* D1, 2252.

Bunner, A. N., Coleman, P. L., Kraushaar, W. L., and McCammon, D. 1972, *Ap. Lett.* 12, 165.
Cameron, A. G. W., and Mock, M. 1967, *Nature* 215, 464.
Catura, R. C., Acton, L. W., Johnson, H. M., and Zaumen, W. T. 1974, *Astrophysical Journal* 190, 521.
Cowsik, R., and Kobetich, E. J. 1972, *Astrophysical Journal* 177, 585.
Cruddace, R., Paresce, F., Bowyer, S., and Lampton, M. 1974, *Astrophysical Journal* 187, 497.
Davies, R. D., and Smith, F. G., eds. 1971, *The Crab Nebula* (IAU Symp. No. 46) Reidel.
Davison, P. J. N., Culhane, J. L., and Morrison, L. V. 1975, *Nature* 253, 610.
Deerenberg, A.J.M., Bleeker, J.A.M., de Korte, P.A.J., Yamashita, K., Tanaka, Y., and Hayakawa, S. 1973, *Nature Phys. Sci.* 244, 4.
Eddington, A. S. 1926, *Internal Constitution of the Stars*, p. 115, Cambridge University Press.
Fabian, A. C., and Sanford, P. W. 1971, *Nature Phys. Sci.* 231, 52.
Fabian, A.C., Zarnecki, J.C., Culhane, J.L., Hawkins, F.J., Peacock, A., Pounds, K.A., and Parkinson, J.H. 1974, *Astrophysical Journal (Lett.)* 189, L59.
Fazio, G. G. 1973, in Stecker and Trombka, *op. cit.*, p. 153.
Felten, J. E. 1970, in Gratton, *op. cit.*, p. 216.
______. 1973, in Bradt and Giacconi, *op. cit.*, p. 258.
______. 1974, in *Planets, Stars and Nebulae Studied with Photopolarimetry*, ed. T. Gehrels, University of Arizona Press, p. 1014.
Felten, J.E., Gould, R.J., Stein, W.A., and Woolf, N.J. 1966, *Astrophysical Journal (Lett.)* 146, 955.
Felten, J.E., and Humphreys, R.M. 1973, *Astrophysical Journal* 181, 543.
Felten, J.E., and Morrison, P. 1966, *Astrophysical Journal* 146, 686.
Felten, J.E., and Rees, M.J. 1969, *Nature* 221, 924.
______. 1972, *Astr. Ap.* 17, 226.
Felten, J.E., Rees, M.J., and Adams, T.F. 1972, *Astr. Ap.* 21, 139.
Field, G. B. 1972, *Ann. Rev. Astr. Ap.* 10, 227.
Fireman, E. L. 1974, *Astrophysical Journal* 187, 57.
Giacconi, R., Gursky, H., Kellogg, E., Levinson, R., Schreier, E., and Tananbaum, H. 1973, *Astrophysical Journal* 184, 227.

Giacconi, R., Murray, S., Gursky, H., Kellogg, E., Schreier, E., Matilsky, T., Koch, D., and Tananbaum, H. 1974, *Astrophysical Journal Suppl*. 27, 37.

Gottlieb, E. W., Wright, E. L., and Liller, W. 1975, *Astrophysical Journal (Lett.)* 195, L33.

Gratton, L., ed. 1970, *Non-Solar X- and Gamma-Ray Astronomy* (IAU Symp. No. 37) Reidel.

Harris, D. E., and Romanishin, W. 1974, *Astrophysical Journal* 188, 209.

Hegyi, D. J., ed. 1973, Sixth Texas Symposium on Relativistic Astrophysics. *Ann. N.Y. Acad. Sci*. 224, 1.

Heitler, W. 1954, *Quantum Theory of Radiation*, 3rd edition p. 250, Clarendon Press.

Illarionov, A.F., and Sunyaev, R.A. 1972, *Sov. Astr. - AJ* 16, 45 (*Astr. Zhur*. 49, 58).

Kardashev, N. S. 1962, *Sov. Astr. - AJ* 6, 317 (*Astr. Phur*. 39, 393).

Karzas, W.J., and Latter, R. 1961, *Astrophysical Journal Suppl*. 6, 167.

Kellogg, E. M. 1971, in Davies and Smith, *op. cit*., p. 42.

______. 1973, in Bradt and Giacconi, *op. cit*., p. 171.

Kellogg, E., Murray, S., Giacconi, R., Tananbaum, H., and Gursky, H. 1973, *Astrophysical Journal (Lett.)* 185, L13.

Lamb, F.K., Pethick, C.J., and Pines, D. 1973, *Astrophysical Journal* 184, 271.

Lampton, J., Margon, B., Bowyer, S., Mahoney, W., and Anderson, K. 1972, *Astrophysical Journal (Lett.)* 171, L45.

Laros, J.G., Matteson, J. L., and Pelling, R.M. 1973, *Astrophysical Journal* 179, 375.

Lyuty, V.M., Shakura, N.I., and Sunyaev, R. A. 1975, *Sov. Astr. - AJ*, in press (*Astr. Zhur*. 51, 000, 1974).

Margon, B. 1974, *Nature* 249, 24.

Margon, B., Bowyer, S., and Stone, R.P.S. 1973, *Astrophysical Journal (Lett.)* 185, L113.

Margon, B., and Ostriker, J.P. 1973, *Astrophysical Journal* 186, 91.

McBreen, B., Ball, S. E., Jr., Campbell, M., Greisen, K., and Koch, D. 1973, *Astrophysical Journal* 184, 571.

Metzger, A.E., Parker, R.H., Gilman, D., Peterson, L.E., and Trombka, J.I. 1974, *Astrophysical Journal (Lett.)* 194, L19.
Moore, W.E., Cordova, R., and Garmire, G.P. 1973, in *Conference Papers*, 13th International Cosmic Ray Conference, 1, 56.
Novick, R., Weisskopf, M.C., Berthelsdorf, R., Linke, R., and Wolff, R.S. 1972, *Astrophysical Journal (Lett.)* 174, L1.
Osmer, P.S., and Hiltner, W.A. 1974, *Astrophysical Journal (Lett.)* 188, L5.
Page, D.N., and Thorne, K.S. 1974, *Astrophysical Journal* 191, 499.
Pringle, J.E., Rees, M.J., and Pacholczyk, A.G. 1973, *Astr. Ap.* 29, 179.
Rees, M.J. 1974, in *Highlights of Astronomy* 3, ed. G. Contopoulos, p. 89, Reidel.
Ricker, G.R., Scheepmaker, A., Ryckman, S.G., Ballintine, J.E., Doty, J.P., Downey, P.M., and Lewin, W.H.G. 1975, *Astrophysical Journal (Lett.)* 197, L83.
Scargle, J.D., and Harlan, E.A. 1970, *Astrophysical Journal (Lett.)* 159, L143.
Schwartz, D., and Gursky, H. 1973, in Stecker and Trombka, *op. cit.*, p. 15.
Setti, G., and Rees, M.J. 1970, in Gratton, *op. cit.*, p. 352.
Seward, F.D., Burginyon, G.A., Grader, R.J., Hill, R.W., and Palmieri, T.M. 1972, *Astrophysical Journal* 178, 131.
Shakura, N.I., and Sunyaev, R.A. 1973, *Astr. Ap.* 24, 337.
Silk, J. 1970, *Spa. Sci. Rev.* 11, 671.
______. 1973, *Ann. Rev. Astr. Ap.* 11, 269.
Stecker, F.W. 1969, *Astrophysical Journal* 157, 507.
______. 1973, in Stecker and Trombka, *op. cit.*, p. 211.
Stecker, F.W., Morgan, D.L., Jr., and Bredekamp, J. 1971, *Phys. Rev. Lett.* 27, 1469.
Stecker, F.W., and Trombka, J.I., eds. 1973, *Gamma-Ray Astrophysics*, National Aeronautics and Space Administration.
Stevens, J.C., Riegler, G.R., and Garmire, G.P. 1973, *Astrophysical Journal* 183, 61.
Strittmatter, P.A., Scott, J., Whelan, J., Wickramasinghe, D.T., and Woolf, N.J. 1973, *Astr. Ap.* 25, 275.

Strong, I.B., Klebesadel, R.W., and Olson, R.A. 1974, *Astrophysical Journal (Lett.)* 188, L1.
Sunyaev, R.A. 1970, *JETP Lett.* 12, 262 (*ZhETF Pis. Red.* 12, 381).
Tananbaum, H., Gursky, H., Kellogg, E.M., Levinson, R., Schreier, E., and Giacconi, R. 1972, *Astrophysical Journal (Lett.)* 174, L143.
Trimble, V., and Rees, M. 1970, *Ap. Lett.* 5, 93.
Tucker, W. 1967, *Astrophysical Journal* 148, 745.
Tucker, W.H. 1970, *Astrophysical Journal* 161, 1161.
Tucker, W.H., and Gould, R.J. 1966, *Astrophysical Journal* 144, 244.
Turland, B.D. 1975, *M.N.R.A.S.* 170, 281.
van den Heuvel, E.P.J., and De Loore, C. 1973, *Astr. Ap.* 25, 387.
Wallerstein. G. 1967, *Ap. Lett.* 1, 31.
Williamson, F.O., Sanders, W.T., Kraushaar, W.L., McCammon, D., Borken, R., and Bunner, A.N. 1974, *Astrophysical Journal (Lett.)* 193, L133.
Woltjer, L. 1958, *Bull. Astr. Insts. Neth.* 14, 39.
_______. 1970, in Gratton, *op. cit.*, p. 208.